The Professional Coach's Assistant;
A Coaching Workbook & Guide

Written by Rev. Dr. Kevin T. Coughlin Ph.D.

KTC Publishing Phase IIC Coaching, LLC

This book is a work of nonfiction.

First Printing

Copyright © 2017 Rev. Dr. Kevin T. Coughlin Ph.D.

KTC Publishing Phase IIC Coaching, LLC

All Rights Reserved.

As permitted under the US Copyright Act of 1976, no part of this publication may be reproduced, distributed, or transmitted in any form or by any means (electronic, mechanical, photocopying, recording, or otherwise, stored in a database or retrieval system, without prior written permission of the copyright holder of this book, except by a reviewer, who may quote brief passages in a review.

The scanning, uploading, and distribution of this book via the Internet or other means without the written permission of the copyright holder and publisher is illegal and punishable by law. Please purchase only authorized electronic editions, and do not participate in or encourage electronic piracy of copyrighted materials. Your support of the author is appreciated.

Printed in the United States of America

ISBN 978-1985731998(paperback)

The Professional Coach's Assistant; A Coaching Workbook and Guide has been used successfully by numerous individuals, Professional Life and Recovery Coaches, residential recovery programs, outpatient programs, aftercare professionals, counselors, therapists, probation officers, ministries, recovery retreats, sponsors, sober companions, and family members to help them to get a deeper understanding of the Professional Coaching process. This amazing workbook and guide has everything a Professional Coach will need to supplement their original training and start coaching clients. For clients, this Workbook and Guide is a great aid for Professional Coaches to achieve success with their clients! Professional Coaches and clients, get your copy today, you will be pleased that you did!

Rev. Dr. Coughlin has been helping to change and save thousands of lives for over two decades. He is a prolific writer, a Best-Selling Author, Award-Winning Poet, speaker, consultant, Master coach, interventionist (CIP), Christian therapist, Pastoral Counselor, and expert on Coaching and addiction. The workbook and guide is tried, tested, and proven as a winner!

This workbook and guide for coaches and their clients should be a tool in every coach's office and briefcase. If you're a coach, and you don't have this workbook, you are missing out! If you're a client, and your coach doesn't use this workbook, you're missing out! Coaches and their clients everywhere will benefit from this tremendous workbook and guide.

Coaching should be life changing, results and client-driven, and solution focused. This workbook and guide will aid coaches everywhere to help their clients to change their perceptions and perspectives and take the action necessary to get results.

PLEASE, VISIT www.theaddiction.expert for other books written and published by Rev. Dr. Kevin T. Coughlin Ph.D., there you can join his mailing list for advanced notice on their next books, training, and live events.

Disclaimer: In this book, the author shares his experience, strength, and hope with readers, this should not be considered advice. All information in this book is for informational and educational purposes, not medical or psychiatric advice or to prescribe the use of any technique as a form of treatment for medical or psychiatric problems without the advice of a physician, psychiatrist, or appropriate licensed professional either directly or indirectly. In the event, you use any of the information in this book for yourself, neither the authors nor the publisher accepts responsibility for your actions.

Introduction

Professional Life and Recovery Coaches have dedicated their lives to helping those who have problems find solutions. Professional Coaching is client and goal driven, where coaches utilize specialized tools, skill sets, and core competencies to help the client to change perspectives and perceptions and find solutions that were unobtainable to them in the past. Professional Coaches do not conduct process work, they leave that for counselors and therapists. Coaches stay focused on today and moving forward. Coaches set a solid foundation with clients through professional standards and ethical guidelines, establishing an agreement with their clients and establishing a presence and relationship based on trust and intimacy. Professional Coaches are expert communicators utilizing active listening, powerful questioning, and direct communication to assist in creating client awareness, goal setting, and designing action plans, and then managing the client's progress while holding them accountable. Coaches demonstrate flexibility, creativity, understanding, and compassion, and celebrate all the victories along the road to success with their clients. Professional recovery coaches partner with clients in a thought-provoking process that is designed to be creative and inspirational so that clients maximize their lives potential and promise. Ultimately coaches set clients up for success!

Table of Contents

Introduction 4
The Professional Recovery Coach's Questionnaire and Forms 1
Anger Log 8
Needs Chart 9
Life Coaching Chart 11
Drug and Alcohol Questionnaire 12
This is the Life Pie Exercise 16
Report Writing 17
Action Plan 18
These are the original Twelve Steps as published by Alcoholics Anonymous 19
Diagram 20
Your Heart's Desire Exercise 21
Finding Your False Beliefs 22
What's Not Wrong in Your Life? 23
Priorities and Rocks? 24
Diagrams 25
Positive and Negative Traits 29
Personal Timeline 30
Bucket List 31
The Box 32
My Gratitude List 33
Who Am I? 34
Write a Good-bye Letter to Your Addiction 35
Goal Setting 36
Celebrate All the Victories! 37
Random Acts of Kindness 38
Hero 39
If I were? 40
Life Influences 41
Treat Yourself Better Exercise 42
Insomnia Due to Anxiety Exercise 43

Self-esteem Collage Board	44
Keeping a Healthy Eating and Exercise Journal	45
My Self-esteem Calendar	46
Confidence Building Exercise	47
The Master Coaching Plan: The Core Competencies	48
The Art of Communication and Powerful Questioning as a Coach	53
The Art of Listening	57
The Coach's Power Pyramid!	59
Legal Forms and Contracts for Professional Coaching	63
Cancelation Policy	63
Coaching Agreement Contract	64
Confidentiality Agreement Contract	65
Contract for Coaching Sessions	69
Contract for Intervention Services	70
Contract for Intervention	71
Random Drug Test Sheet	73
Drug Testing Consent Form	74
Informed Consent	78
Medical Record Release Form	79
Medical Release Form	80
Minors Guardianship	83
Payment Agreement	84
Professional Coaching Contract	90
Professional Services Agreement	91
Sober Coach/ Sober Companion Live-In Agreement	97
Financial Agreement	98
Each Coach's Responsibility State to State	99
Notes	100
Rev. Dr. Kevin T. Coughlin Ph.D. Publication Credits	1
About the Author	6
Follow Rev. Kev. on Social Media	7

The Professional Recovery Coach's Questionnaire and Forms

Coach introduction: good morning, afternoon, or evening, my name is _____ thank you for calling/ coming in today. I appreciate your time.

Do you have any questions before we get started? If yes, answer briefly. If no, move on.

(Questions)

Last name: _____, Middle: _____, First: _____
What is your date of birth? ____/____/_____   Age: _____
Address: _____   _____   _____   _____
Phone Number: (____)-_____-_____ Cell Number: (____)-_____-_____
Emergency Contact: name: _____ Phone Number: (____)-_____-_____   Relationship: _____ 2nd Number: (____)-_____-_____
How did you hear about my company? __Friend, __Website, __Ad, __Article, __Referral   Referred by: _____.
What is your main goal in coming to see me today? _____   _____.
What have you specifically done about the problem yourself?   _____   _____.

Referral Resources (Circle all that apply)

Self-referral	Court	Therapist
Church/clergy	Corrections	Family Friend
Treatment Provider	Family	Recovery Support Services
School	Child Welfare	Employer
Physician	Counselor	Other

How do you feel about being here today? (Circle all that apply)

Angry	Uncertain	Hopeful
Resentful	Reserved	Happy
Anxious	Resigned	Excited
Fearful	Afraid	Determined
Scared	Confused	Ready

What would you like to accomplish through coaching? (Circle all that apply)

Maintain Sobriety	Get Employment	Obtain Housing
Recovery Support	Child Custody	Food/Clothing
Meet Legal Requirements	Improve Relationships	Avoid Jail/Prison
Anger Management	Get Partner/Spouse Back	Recovery Network
Become More Spiritual	Improve Life	Become Better Person
Understand Addiction	Understand Recovery	Change Thinking
More Service Work	Get Sponsor	Change How I Live
Go to AA/NA meetings	Pray/Meditate More	Other

Please list you most important goals in order of importance to you.

1.	2.	3.	4.
5.	6.	7.	8.

On a scale of 1-5 with (5) being very confident and (1) being not confident, how confident are you that you will be able to achieve each one of your listed goals?

Please place the corresponding number in each box.

1.	2.	3.	4.
5.	6.	7.	8.

On the same scale, how ready are you to start working on your goals today?

Please place the corresponding number in each box.

1.	2.	3.	4.
5.	6.	7.	8.

Please circle all that apply to you. If you have an emergency and need help, call 911.

1. Not Safe at Home	2. Suicidal	3. Abused	4. Risk of Relapse
5. Homeless	6. No Food	7. Mental Health	8. No Medication
9. Sexual Abuse	10. Special Needs	11. Not Safe In Neighborhood	12. Need Medical Attention

EMERGENCIES CALL 911
National Suicide Prevention Lifeline
1-800-273-8255

Please check the boxes that apply to you for transportation.

	Valid License to Drive		License Suspended		Need Help to Get License to Drive		Need Funds to Get License
	Own a Car		Don't Have a Car		Family Has Car		Want to Buy a Used Car
	Public Transportation		Don't Have a Bus Pass		Have a Bus Pass		Need Help With Funds
	Ride a Bike		Need a Bike		Bike Needs Repair		Need Funds to Repair Bike
	Walk to Where I Need to Go		Disabled Can't Walk		Need Comfortable Walking Shoes		No Funds for Shoes

Please circle any special transportation needs you may have.

Special Needs Hearing Impairment	Special Needs Visual Impairment	Need Wheelchair/Handicap Access
Special Needs Due to Physical Mobility Restrictions	Walk with Assistance Cane/ Walker	Other (specify)

Please circle anything that relates to your current employment status.

Are you employed?	Full-time student	Work part-time irregular hours
Laid off	Someone supports me	In the military
Terminated	Retired	Volunteer work only
Quit job	Disabled	Like my job/ don't like job
Out of work over ninety days	Can't find wok/legal problems	Looking for a new job
Looking for work	Just got out of jail	I have more than 1 job
Want to work/can't find job	Work full-time 35+ hours	My job is good for recovery, my job is bad for my recovery
Choose not to work	Work part-time reg. hours	Job does not affect my recovery

Circle all job skills that apply to you.

1. Child Care	2. Trucking	3. Business	4. Professional
5. Customer Service	6. Health Care	7. Sales	8. Office
9. Trade	10. Landscaping	11. Supervision	12. Other

Think of the job skills that you have. List them in order of what is your best skill from 1-12, 1 being your best, 12 being your worst.

1.	2.	3.	4.
5.	6.	7.	8.
9.	10.	11.	12.

List eight skills that you would like to develop and get more experience in.

1.	2.	3.	4.
5.	6.	7.	8.

Please circle all the responsibilities that you have outside of work.

Child Care	Care for Disabled Family	Housework/ Chores	School and Homework
Care for Elderly Family	Mandatory Reporting Probation	Report to Case Manager	Take Care of Pets
Mandatory AA/NA Meetings	Report to Out Patient	Step Work with Sponsor	Parole Requirements

Do you think it would benefit you to get some help with employment/work related matters? Y/N

Circle the type of help that you would like.

Bettor Job	Improve Interviewing Skills	Vocational Assessment
Developing Job Skills	Develop Resume	Help Finding Work
Disability Evaluation	Promotions	Maintain Job
Disability Work Rehabilitation	Interviewing Skills Arrange Interviews	Barriers Related to Felony Convictions

Please circle your education level.

High School Diploma	College No Diploma	Bachelor's BA, BUS
High School didn't finish	College 1 year, 3 years	Master's Degree MA, MS
High School/GED Voc/Tech Diploma after High School	College Associates Degree AA, AS	Doctorate multiple doctorates

Please list certifications and training.

Are you a student now or in any specialized training? Y/N

Please circle any item that you're interested in getting help with.

Earn GED	Academic Counseling	Tutoring
Literacy Training	Aptitude/Achievement Tests	Grants/Loans for School
Returning to school additional training	Technical/Vocational Training	Applying to schools other: _____

Please circle who you live with.

With Spouse/Domestic Partner and Children	With Spouse/Domestic Partner	With Child/Children	With Parent/Parents
With Other Family	With Friends	Alone	Controlled
Homeless	Temporary	Group Home	Other: _____

Please circle all that apply to your living situation. If any question makes you uncomfortable, leave it out.

People Support Recovery	Threatening	My neighborhood is safe
People are in Recovery	Intimidating	Neighborhood not safe
People do not Support	Verbally Abusive	Neighborhood is dangerous
People keep drugs in house	Physically Abusive	Not dangerous
People keep alcohol in house	Sexually Abusive	Good for recovery
People use drugs and alcohol	I am safe in house	Bad for recovery
People sell drugs	I am not safe in house	Would like to move
People discourage my recovery	Family safe, family not safe	Would like to stay

Circle any of these you may be interested in.

Emergency Housing	Temp Housing	Recovery Home	Stable Housing
Supported independent living	Help finding Subsidized Housing	Oxford House, other clean and Sober Housing	Housing barriers related to felons

Have you experienced any legal trouble? If yes, circle all that apply to you.

Ever been arrested	DUI arrest	PTA	Parole
Convicted felony	DUI conviction	AR program/ fines	Probation
Misdemeanor	Jail time/ prison time	Child Support	Active Warrants
Charges dropped	Community Service hours due	Public Defender/ paid attorney	Awaiting trial

Please describe your mental health history below.

Mental Health diagnosis Y/N?	What is your MH diagnosis?	Are you prescribed any medications for MH issues? Y/N
Have you ever been hospitalized because of MH reason? Y/N		II yes, what are they? _____
If yes, where and why?	Have you ever had detox complications? Y/N	Have you ever had treatment for addiction? Y/N?
Where:	If yes, what?	If yes, where?
Why:		

Circle the statement that best describes your current situation.

I don't have an addiction problem.	I'm in recovery and have been drug and alcohol-free for over 1 year.	I'm in early recovery and have been clean and sober for ninety days.
I haven't used in one week.	I have used in the last week.	I'm actively using.
I want to stop using but can't stop.	I don't want to stop.	I'm court ordered to stop.

Please answer the questions below as they relate to you.

Do you have a recovery plan? Y/N	Do you want help to create a recovery action plan? Y/N
Do you want help to update your recovery plan? Y/N	Who if anyone is helping you meet current recovery goals? sponsor, case manager, recovery coach, recovery support services, or other:
What person do you look to when you need help? _____	Are you interested in working with a trained professional who can help you to reach your recovery goals and solutions? Y/N

I have listed eight sober activities that you could do; please list eight of your own.

1. Writing/poetry	5. Lifting weights	9.	13.
2. Walking	6. Bowling	10.	14.
3. Music	7. Hiking	11.	15.
4. Art	8. Gardening	12.	16.

Please circle all items that apply to you that you may need assistance with.

Bankruptcy 7,13	Alimony	Debt	Discrimination
Criminal Record	Child Support	No Health Insurance	Disability
Immigration Status	No income	Paying for meds	Getting divorced
Child Custody	Child Custody Pay	Alimony Pay	Probation problems
Legal Defense	Need food stamps	Need clothing	Need personal items
Need detox	Need treatment	Need housing paid	Need heating paid
Need diapers	Need a phone	Need money	Need child care
Need help reading	Can't speak English	Need a mentor	Need help

Anger Log

Date	Angry at?	Why I am angry and what I did about it? (Positive or Negative behavior.)	P/N

Needs Chart

Client	Date	Time	Agency	Address	Trans
John Doe	1/1/14	1500	Parole	Main St. Allentown	Needs

Chairperson sign	Date	NA/AA	Signature	Date	Agency	Time
Signed by Chairperson	1/1/14	NA	Officer Blue Signs	1/1/14	State Parole	1600
"Ditto"	1/15/14	AA				
"Ditto"	1/16/14	AA				
'Ditto"	1/17/14	AA				
"Ditto"	1/18/14	NA				
"Ditto"	1/19/14	NA				

Check off areas that pertain to your client. Appointment dates and time are written on top.

Client	Agency	Date	Time	Agency	Date	Time	Agency	Date	Time
J. Doe	Parole	1/1/14	1500						
X	Parole			Probation		Detox		Volunteer		Minor
X	Fines due			Therapy		Res. Treat.		Counseling	X	Transport
	Dental			Social Ser.		I.O.P.		Financial		Sponsor
	Doctor			Work		Rec Coach		Special Needs		Religion
	Housing			Picsy.		School		Citizenship		Church
X	AA/NA/Rec.			Drive License		Family		Anger Manage		Memberships
	Child Support			Drug Court		Children		Intervention		Insurance
	Food			Clothing		Custody		Drug Testing		Surgery
	Medications			Legal		Hygiene		Court Order		Mental Health

Life Coaching Chart

- 1. There are ninety-eight hours in a week at seven days some week fourteen hours per day.
- 2. How much time does the client spent on the following areas each week.
- 3. Health. How important is their health to them?
- 4. Wealth. What does wealthy mean to client?
- 5. Family. How important is family to them?
- 6. Relationships. How important are relationships and who are the important people?
- 7. Job/Career. What's important in client's work?
- 8. Spiritual. How important is spiritual growth?
- 9. Playtime. What does client do for fun?
- 10. Contribution. World contribution. How important?
- 11. What areas do you lack in? How important?

In each category what is client's current situation, how much time each week, and what is the client's future goals? You can make this into a chart if you choose.			
Health	1-2-3-4-5-6-7-8-9-10	Wky hrs:	
Wealth	1-2-3-4-5-6-7-8-9-10	Wky hrs:	
Relationships	1-2-3-4-5-6-7-8-9-10	Wky hrs:	
Family	1-2-3-4-5-6-7-8-9-10	Wky hrs:	
Career/Job	1-2-3-4-5-6-7-8-9-10	Wky hrs:	
Spiritual	1-2-3-4-5-6-7-8-9-10	Wky hrs:	
Playtime	1-2-3-4-5-6-7-8-9-10	Wky hrs:	
World	1-2-3-4-5-6-7-8-9-10	Wky hrs:	
Category	**importance 10 most**	**weekly hours**	**future goals**
Health	7	1	Want to improve
Wealth	10 being able	40	More wealth
Relationships	7	5 Out of	Improve
Family	8	7 Balance!	Improve
Career	10	45	Improve
Spiritual	10	0	Get more
Playtime	2	0	More
World	7	Overlaps	Improve

Drug and Alcohol Questionnaire

Client name: **Coach:**   **Date:**

Circle any drug that you have used, and explain first and last use, and if ever a problem:
Alcohol
Barbiturates
Sleeping pills
Benzodiazepines
Caffeine
Cocaine
Crack
Ecstasy (MDMA)
Ephedra
Gasoline
Glue
Heroin
Other Inhalants
LSD
Marijuana or Hashish
Methadone
Methamphetamine
Mescaline
Mushrooms
Nicotine
Nitrous Oxide
Opiates (pain pills)
Opium
PCP
Peyote
Poppers
Prescription Drugs
Psilocybin
Quaaludes
Seconaol (Reds)
Speedballs
Steroids
Tino (Yellows)

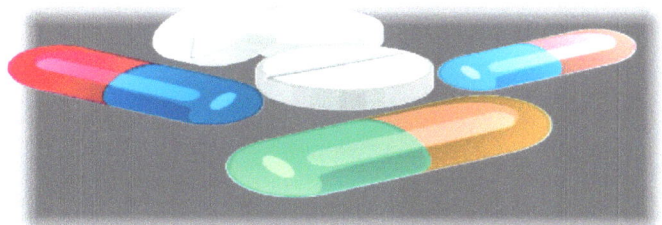

Please put a circle around any of the drugs above that you feel you are addicted to or dependent upon.

How did you get started using drugs/alcohol?
When you consume alcohol, what do you usually drink?
How many drinks do you usually have per day or per week?
How much (name of drug) do you usually have per day or per week?
How have you ingested (the drug)?
Swallow/Smoke/Sniff/Inject/Mix with other
What is the best thing about getting high?
What is your favorite thing to do when drinking or using drugs?
Are there any times you tend to use these substances less?
Are there any times you have successfully stopped?
How much do you spend each week on your drugs/alcohol?
Do you usually drink/use drugs alone or with others?
At home or elsewhere?
What time of day do you usually start using drugs/drinking?
Is there a pattern to your use?
What effects does drinking/using drugs have on your feelings and emotions?

Do you or have you ever experienced any physical symptoms when you try to stop drinking or use drugs?

If so, which ones? Shakes/tremors, sweating, seizures, continuous vomiting, sleeplessness, disorientation, hallucinations, depression, hypersomnia, increased appetite, other:
Do you gamble when you drink or use drugs?
Is your gambling out of control or excessive?
Have you ever had an eating disorder (bulimia, anorexia, obesity)?

Which family members have had a drug or alcohol problem (circle)?

How were you affected by your family member's drug abuse?
Does anyone in your household use drugs or drink?
If so, who?
Do most of your friends drink or use illicit drugs?

Please circle any problems that have persisted following your use of drugs or alcohol:

Hepatitis or liver problems, persistent cough, hallucinations, strange thoughts, congestion or wheezing, heart problems, depression, mania. Other:

Please circle any social or relationship problems that have resulted from your use of alcohol or drugs:

Arguments with spouse or partner, thrown out of house, social isolation, arguments with parents or siblings, loss of friends, spouse or partner left you, other:

Please circle any job or financial problems caused or worsened by your use of drugs or alcohol:

Lost a job, less productive at work, behind in paying bills, late to work, in debt, bankruptcy, foreclosure, repossession, missed days at work, missed opportunities for raise or promotion, other:

Please circle any problems caused or worsened by use of alcohol or drugs:

Arrest for possession, forging prescriptions, assault, embezzlement, forgery, selling drugs, driving under the influence, arson, sexual assault, hate crimes, homicide, theft or robbery.

Have you ever attended a twelve-step program?

Have you ever gone to an outpatient program for drugs or alcohol?

Have you ever been in an inpatient facility for drugs or alcohol?

Have you ever used a prescription medication to abstain from drinking or using drugs?

Have you ever had a drug overdose or alcohol poisoning?

Have you ever attempted suicide while intoxicated or using?

What is the longest period of not using you have had to date?

How have you stayed clean and sober so far?

What caused you to want to stop using?

What do you think the result will be if you keep using?

Please write a T if True and an F is False at the end of each question.

1. I drink/use drugs when I feel anxious.
2. I often try to hide or minimize my drinking/drug use.
3. Many of my friends drink or use illicit drugs.
4. I have broken the law to support my habit.
5. I would never consider going to a twelve-step program.
6. Drinking or using drugs has never really caused me any problems.
7. I have tried to stop using drugs/drinking in the past.
8. I drink/use drugs when I feel depressed.
9. When I drink, I usually get drunk.
10. I feel more confident when I drink or use drugs.
11. Sometimes I use drugs or drink in the morning.
12. Friends or family have told me I should stop drinking or using drugs.
13. I spend too much time thinking about drinking or using drugs.
14. I become very anxious if I am unable to have a drink or do drugs.
15. I have never stolen in order to buy drugs or alcohol.
16. I am an alcoholic.
17. I am a drug addict.
18. I have experienced the need to use more drugs to get the effect I had the first time I used them.
19. If I stopped using drugs or drinking, I would lose many of my friends.
20. I am not a religious person.
21. I think better when I have a few drinks or use drugs.
22. I think I have a problem with sexual or gambling addiction.
23. Drinking or using drugs helps me forget about my problems and relax.
24. I have never used drugs and alcohol at the same time.
25. I have sometimes alternated taking uppers and downers.

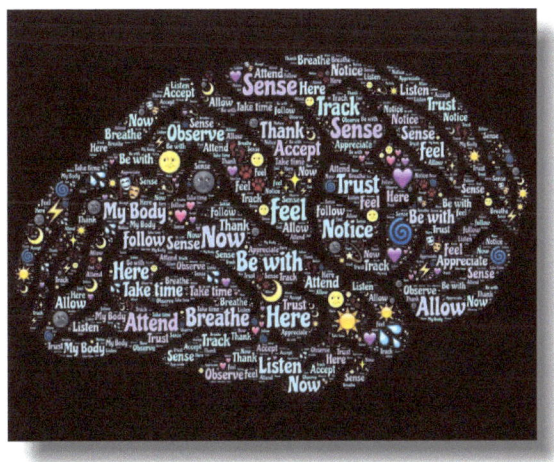

Have you ever experienced any of the following symptoms when you use drugs or alcohol? (circle)

Seizures
Blackouts
Hallucinations
Paranoia
Personality changes
Decreased need for sleep/ Increased aggression/Severe weight loss/ Ulcers/other stomach problems/Headaches/Excessive bleeding/ Sinus problems /Heart palpitations/Suicidal thoughts/Panic attacks/Memory problems/Depression/ Other:

This is the Life Pie Exercise

Divide your pie into eight even slices
at the crust label each of the eight slices
health, wealth, relationships, family, career/job, spiritual, playtime, world
then for each slice of pie divide it into tenths from the center of the pie
the center would be one, then two, half way would be five, the crust would be ten.
ask you client how important each area is in their life on a scale of one through ten
One being the least important and ten being the most important
have them shade the area around the pie that represents the important areas for center outward.

Report Writing

The A, B, C s of Report Writing, be accurate, brief, and clear. Then the five Ws, Who, What, Why, When, Where, and How. When you write a report, an article, or an essay, don't try to sound important using big words. The best writers use words that an eighth grader can understand. Don't try to be someone that you're not, or sound like someone that you're not. Remember, as life coaches, we don't do process work. We go from today forward, never into the past. Let the therapists and Doctors do that! Never make a diagnosis! If you have to send a report to the court, probation, or parole, you can state, it is of my opinion that John needs long-term care. The key words are "it is of my opinion". You don't want to be in a situation where you get accused of diagnosing a client without the credentials or license to do so. Remember confidentiality! If you fall under HIPAA compliance guidelines, remember you have to know who a report is for and have a release to release any information that is protected by law.

Action Planning:

- ❖ Action plans need to be specific.
- ❖ Identify Goals and Objectives.
- ❖ How do you get from point A. to point B.?
- ❖ Who is going to assist you?
- ❖ Why do you need to do this action?
- ❖ Where are you going to do this action?
- ❖ What is the plan?
- ❖ Who will follow up?
- ❖ On what timetable, be specific?

Example:	
Goals and Objectives:	To Improve Communication with coworkers
How:	Clarify Expectations, Meetings
Why:	To Improve Teamwork and Consistency
Who:	John Clarify Expectations, Everyone at Meetings
Where:	XYZ Corp. Meeting Room
When:	Friday, August 22, 2014, 4PM.
Follow Up:	John and Coach
Details of the plan:	Keep Notes, Memos, Daily Info Log, Meeting Minutes, Training for Employees, Employee Manual.

Action Plan

Goals and Objectives:	Specific Dates:

Action:			Specific Dates:

Follow-Up Plan:		Specific Dates:

- ➢ What, how, when, who, why, what? Be specific!
- ➢ Client sets goals and objectives. Coach and client work on an action plan.
- ➢ The plan may have to be adjusted is something is not working, be flexible.

These are the original Twelve Steps as published by Alcoholics Anonymous

1. We admitted we were powerless over alcohol - that our lives had become unmanageable.
2. Came to believe that a Power greater than ourselves could restore us to sanity.
3. Made a decision to turn our will and our lives over to the care of God as we understood Him.
4. Made a searching and fearless moral inventory of ourselves.
5. Admitted to God, to ourselves, and to another human being the exact nature of our wrongs.
6. Were entirely ready to have God remove all these defects of character.
7. Humbly asked Him to remove our shortcomings.
8. Made a list of all persons we had harmed, and became willing to make amends to them all.
9. Made direct amends to such people wherever possible, except when to do so would injure them or others.
10. Continued to take personal inventory, and when we were wrong, promptly admitted it.
11. Sought through prayer and meditation to improve our conscious contact with God as we understood Him, praying only for knowledge of His will for us and the power to carry that out.
12. Having had a spiritual awakening as the result of these steps, we tried to carry this message to alcoholics, and to practice these principles in all our affairs.

Step 1 Gives us the problem	Step 7 Action Step
Step 2 Gives us the solution	Step 8 action Step
Step 3 A decision to live in the solution	Step 9 Action Step
Step 4 Action Step	Step 10 Growth & Maintenance Step
Step 5 Action Step	Step 11 Growth & Maintenance step
Step 6 Action Step	Step 12 Growth & Maintenance step

The Twelve Steps in Their Simplest Form.

Diagram

Below is an example of addiction. It starts with the big lie that allows addicts to use again despite all the negative consequences that have happened as a result of using, this is centered in the mind. Then the addict puts the chemicals into their body and they lose the power of choice. This then causes hopelessness. The ends to addiction are described in the bottom diagram.

THE DISEASE OF THE MIND THE MENTAL OBSESSION 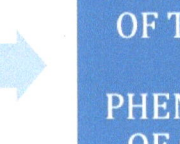 **THE DISEASE OF THE BODY THE PHENOMENON OF CRAVING**  **AS A RESULT THE MALADY OF THE SPIRIT OR HOPELESS CONDITION**

Your Heart's Desire Exercise

Do you feel stressed?
Do you find yourself always acting out of obligation?
Do you know what you really want?
Part of growing up is learning to put the needs of others ahead of ourselves.
When we stop listening to our hearts for a long period of time, it becomes difficult to recognize our heart's voice and life becomes boring!
To recognize our heart's voice once again, we must ask ourselfs what we are feeling when we are stressed and off center and then say to self:
"I would like _____?"
If the answer is practical and possible, then go ahead and do it!
Visit a friend, go for coffee, go to a movie, go out for dinner with friends.
If the impulse is not realistic, like quiting your job, be careful!
Try doing one spontaneous thing every day!
Ask yourself, what would I have done when I was seven?
Start with small baby steps!

Finding Your False Beliefs

Do you ever feel like your life is in a rut, stuck because of unhealthy habits causing false beliefs and unconscious actions?

Complete the sentences below:

1. When under pressure I _____.
2. When _____ happens, I stress out and feel like _____.
3. I often feel guilty about _____.
4. My greatest weakness is _____.
5. I'm always trying to stop _____ from happening.
6. When the unexpected happens I _____.
7. I usually try to _____.
8. The biggest obstacle in my life that stops me from self-love or approval is _____.
9. The thing that I am most afraid of is _____.
10. The thing that drives most of my behavior is _____.
11. I seek approval from _____ always, mostly, usually, occasionally.
12. My most uncomfortable negative emotion is feeling _____.
13. The feeling I dislike the most is _____.
14. I really need to learn to _____.

Now that you have identified your false beliefs go back and re-do the exercise writing how you would like to be.

What's Not Wrong in Your Life?

Take a piece of paper and draw line down the middle so that you have two separate columns.

On one side list everything that is wrong in your life, make sure you get it all out, everything!

On the other side list everything that is right in your life including everything that is not wrong, everything!

When you have completed the two lists, compare them side by side.

Which list is of the greatest use and value to you? Which list serves you to focus on, what's wrong, or what's not wrong?

When you have decided which list is best for you to focus on, discard the other list that you don't wish to focus on, throw it away, bury it, burn it, make it a memorable event!

Try reading the remaining list every day for the next seven to thirty days and see what happens!

Priorities and Rocks?

List the five main things you spend most of your time on every day?
1. _____
2. _____
3. _____
4. _____
5. _____

Which one of the five takes the most of your time and energy?
1. _____

What are really the three top priorities in your life today?
1. _____
2. _____
3. _____

Out of the three which one is the most important of all?
1. _____

If you were to prioritize your responsibilities in order of size of rocks that would fit into a blender first, then pebbles to fill in between the rocks next, and then grains of sand to fill in between the pebbles lastly, how would you prioritize to get as much in the blender to fill it as you could?

Think about it if you just put rocks in the blender it wouldn't be filled, if you added pebbles between the rocks it still wouldn't be filled, but when you added the sand between the pebbles and the rocks, you could fill the entire space.

It's all about first things first and prioritizing so that you can get the most quality out of a day. Another term people use is time management to use your time in the most effective manner to get the most done with the most quality production.

Remember that there are only twenty-four hours in a day and there is always another day to go back to a project. Balance is another key factor in managing your life.

Diagrams

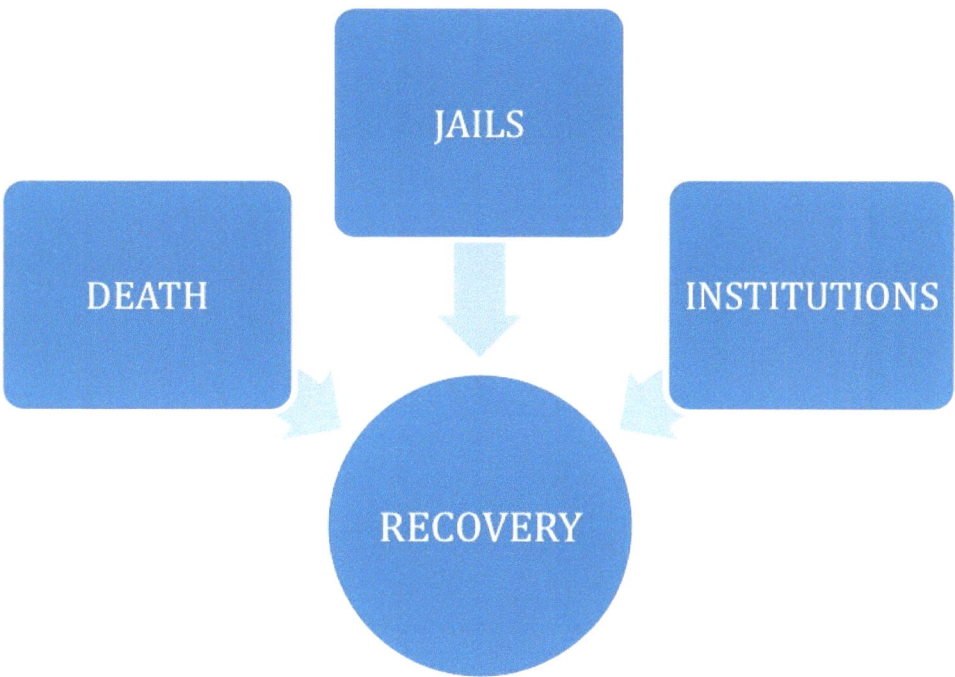

Assume that the family does not know what addiction is. Explain addiction in its simplest form.

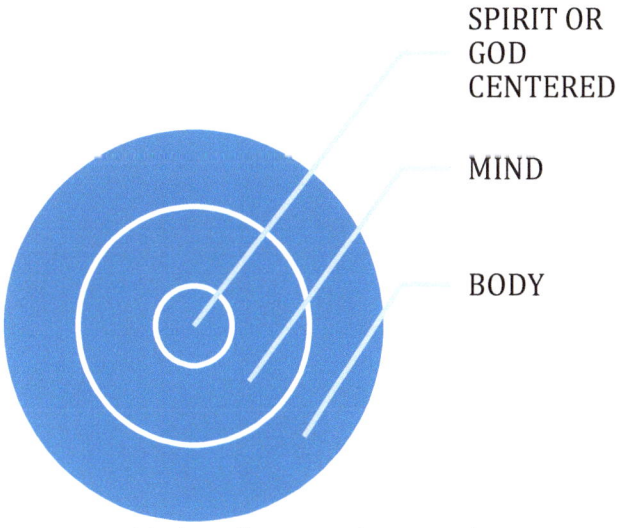

SPIRIT OR GOD CENTERED

MIND

BODY

<u>This is a diagram of a normal person.</u>

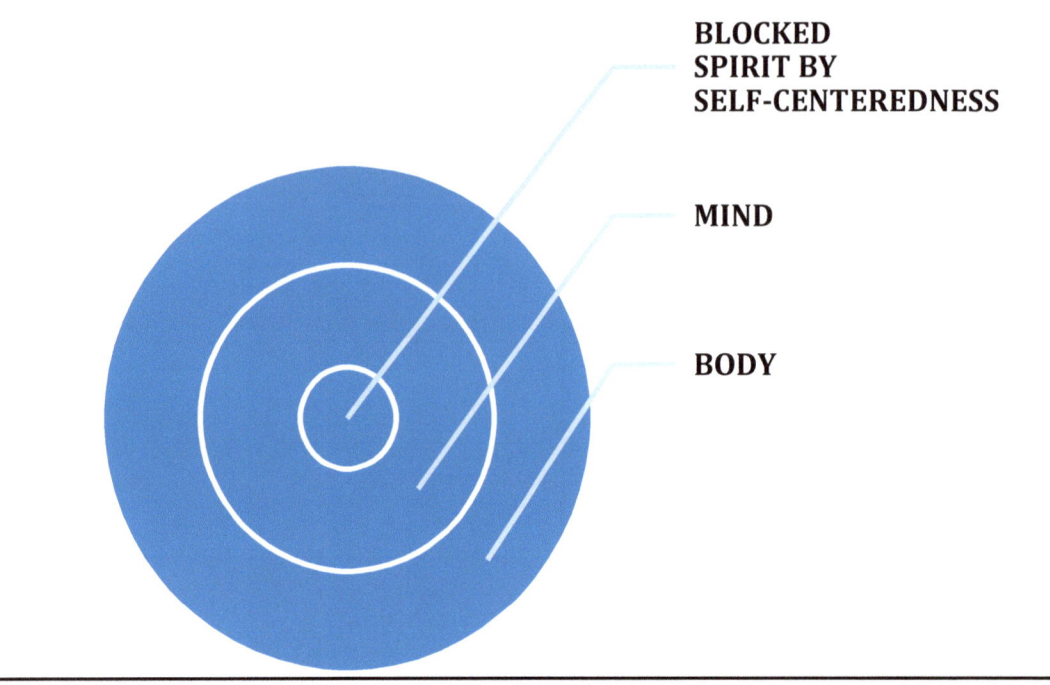

This is a diagram of an addict or alcoholic.
Because the spirit is blocked in the addicted person, the basic instincts become out of balance:

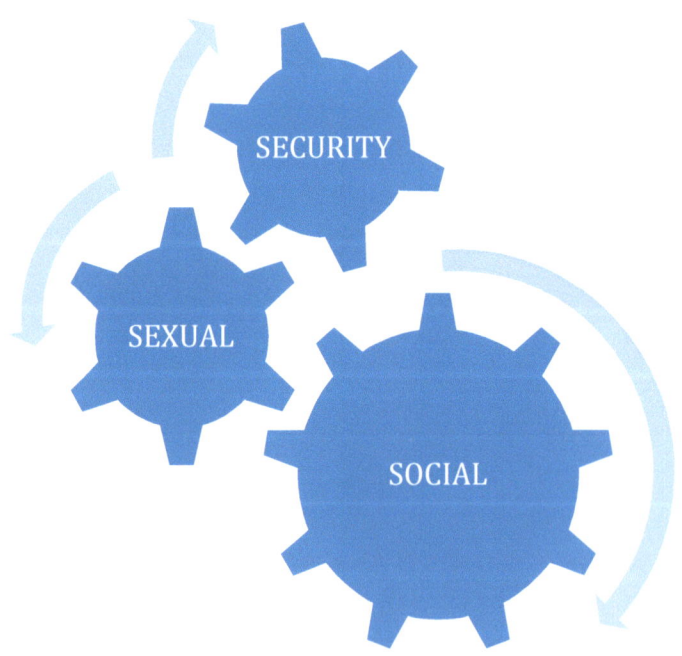

Because the basic instincts are out of balance the addict has fears, harms, and (resentments which are a form of anger.)

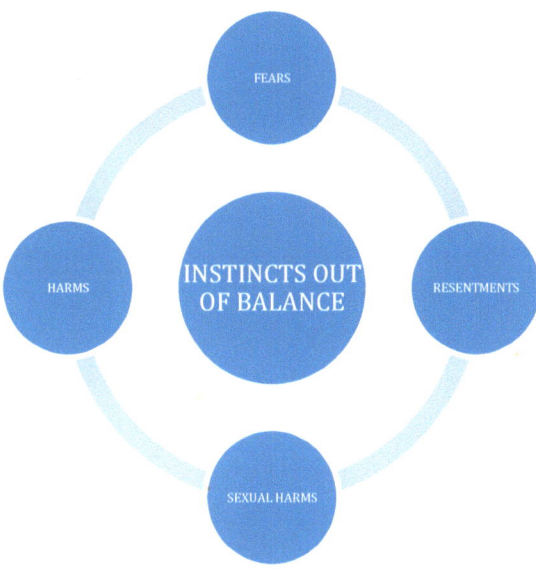

Influence and Trust

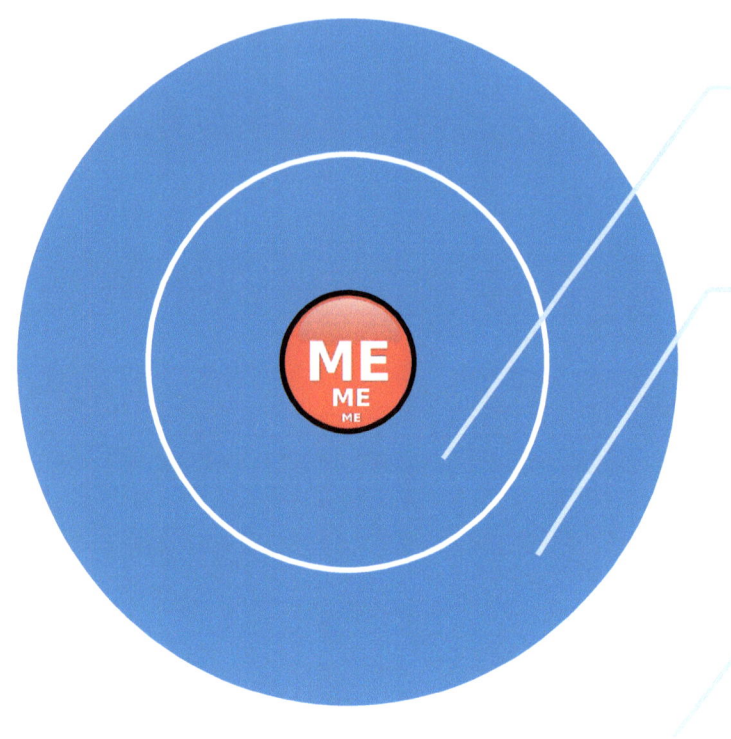

LIST THE TOXIC PEOPLE.

PEOPLE CURRENTLY CLOSE TO YOU.

PEOPLE IN YOUR LIFE IN THE LAST 6 MONTHS.

Last six months	Current	Toxic People
_____	_____	_____
_____	_____	_____
_____	_____	_____
_____	_____	_____
_____	_____	_____

Positive and Negative Traits

+ .... **-**

_____	_____
_____	_____
_____	_____
_____	_____
_____	_____
_____	_____
_____	_____
_____	_____

List your positive traits on the left and your negative traits on the right
List any traits that you're not sure about in the middle.

Personal Timeline

Please write all positive life events above the line and negative events below the line.

Age5____/_____/____/____/____/____/____/____/____/___NOW

Bucket List

Things I want to Accomplish During My Life

1._____
2._____
3._____
4._____
5._____
6._____
7._____
8._____
9._____
10._____

Ideas:
Skydive
Swim with Dolphins
Buy a House
Fall in Love
Be A Champion at Something
Save a Rescue Dog

The Box

This is a group activity.
You can use any size box with a mirror placed inside the box.
Anyone who looks inside the box will see their own reflection.
The groups start with the facilitator asking the group.
"Who do you think is the most awesome person in the World?"
People will usually shout out different celebrity names.
You pass the box around and tell the group as they look in the box they will see the most awesome person in the world!
The point of the group is to show everyone that each and every person is awesome in their own way because of their uniqueness.
You should be prepared for lots of smiles!

My Gratitude List

The people, places, and things that I am grateful for today are:

1. _____
2. _____
3. _____
4. _____
5. _____
6. _____
7. _____
8. _____
9. _____
10. _____
11. _____
12. _____
13. _____
14. _____
15. _____
16. _____
17. _____
18. _____
19. _____
20. _____

Who Am I?

1. _____
2. _____
3. _____
4. _____
5. _____
6. _____
7. _____
8. _____
9. _____
10. _____
11. _____
12. _____

I AM _____.

I

Write a Good-bye Letter to Your Addiction

_____/_____/_____

Dear Addiction:

Sincerely,

Goal Setting

The client always sets the goal!

Okay, so you're stuck and can't figure out what your goal should be.
Your coach can help you through powerful questioning and active listening.

1. Coach asks, "What is the most important problem for you to solve in your life today?"

2. Coach asks, "What will happen if you don't solve that problem today?"

3. Client should now answer both questions, while the coach listens to the client's answers.

4. The client should be able to figure out their goal based on these two great coaching questions.

Remember coach, use open-ended, powerful questions here!

This should help to remove the block that the client has in setting the goal that they need to reach a solution in their life.

Celebrate All the Victories!

It doesn't matter if the progress is big or small, celebrate the wins.

You don't have to throw a giant party, but acknowledge the progress in some positive way.

Think of ten ways the coach and client can celebrate progress together. Consider the first five as smaller wins and the next five as the "big wins!"

Celebrating Victories

1._____
2._____
3._____
4._____
5._____

Bigger Wins!

6._____
7._____
8._____
9._____
10._____

Random Acts of Kindness

Think of one act of kindness that you can do without anyone but you knowing each day for the next seven days. Write each act for each day down below and discuss with your coach at the end of the week.

Monday's Act was _____.

Tuesday's Act was _____.

Wednesday's Act was _____.

Thursday's Act was _____.

Friday's Act was _____.

Saturday's Act was _____.

Sunday's Act was _____.

At the end of the week discuss your random acts of kindness and how it made you feel with your coach.

Hero

My hero is _____.

The positive traits I like about my hero are:

1. _____
2. _____
3. _____
4. _____
5. _____

The positive traits I like about myself are:

1. _____
2. _____
3. _____
4. _____
5. _____

The things I need to improve on are:

1. _____
2. _____
3. _____
4. _____
5. _____

If I were?

> Think about what famous person, super Hero, or Cartoon Character you think you are most like and why?

Write down your answer: _____.

Why?
_____
_____
_____
_____
_____
_____
_____
_____
_____
_____
_____
_____
_____.

If I were?

Life Influences

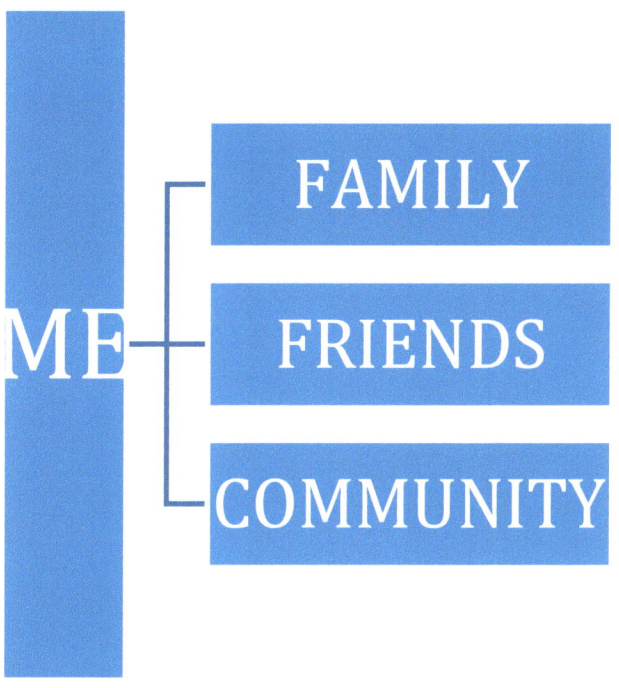

Who made the biggest impact in your life?

Family Members: _____
_____.

Friends: _____
_____.

Community: _____
_____.

Comments: _____
_____.

Treat Yourself Better Exercise

How well do you treat yourself?

Do you think that you deserve good things, fun times, are you nice to yourself?

A great way to improve self-esteem is to learn to love yourself, appreciate yourself, and do things that will make you happy!

You are going to plan a "you" day, where you will be very kind and nice to yourself! You get to do what you want to do, it's your day! Plan fun things that you love to do, cater to yourself, don't hold back, you're worth it!

Plan out your day.
Here is an example:

Breakfast at your favorite diner
A massage at the Spa
Shopping for a new outfit
Visiting an old friend
Lunch with your friend at a favorite restaurant
An afternoon movie
A bubble bath at home
Reading a favorite book by the fire

Be great to yourself, it's your day!

Insomnia Due to Anxiety Exercise

Take control of your anxiety!
You're awake anyway so get up for ten minutes.
Make a list of the top ten tasks that you need to accomplish.
Make sure that you didn't forget anything!

My Ten Minute Take Control List

1. _____
2. _____
3. _____
4. _____
5. _____
6. _____
7. _____
8. _____
9. _____
10. _____

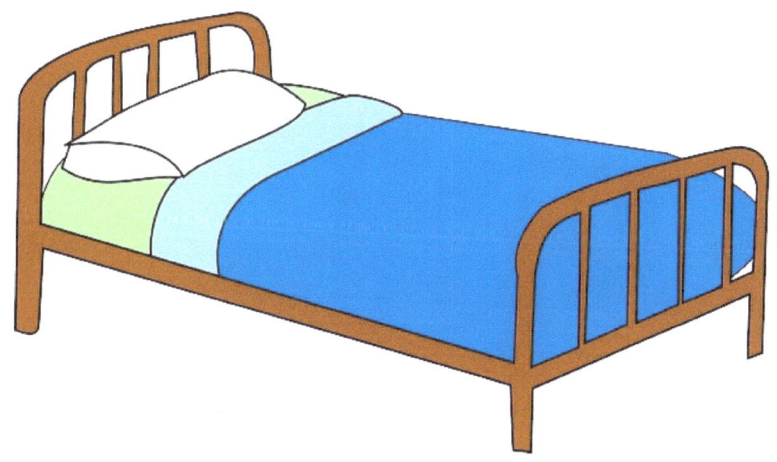

Self-esteem Collage Board

Many people forget about their dreams, goals, hopes, abilities, and talents! Some people let other's block them with negative and hurtful comments.

This exercise is to remind you of who you really are and where you're really headed!

You will be making a value collage to hang on your wall.
Purchase a large poster board
Get together a bunch of magazines to go through
When you look through the magazines, search for pictures that represent your abilities, talents, goals, etc.
This will help you remember who you really are and where you really are headed.

My Power-Lifting Coach used to tell the team, "If you can see it, you can be it!" – Coach Larry P.

Keeping a Healthy Eating and Exercise Journal

Diets don't work!

Healthier eating is about making smarter choices, healthier choices each meal.
You must think about your age, your current health and living situation, access to foods, culture, and traditions.
Every choice you make towards foods and beverages matter.
Start out with small changes until they become healthy habits.
Focus on the amount you put on your plate, the variety, and the nutrition.
Watch for less saturated fat, sodium, and added sugars.
Rules apply to food and the beverages that you consume.
You want to eat the right number of calories for your body according to age, weight, height, sex, and activity level. Your Doctor can help you with this.
Beverage and food choices should be made from all five food groups.

When you make up your meal plate, make half your plate vegetables and fruits (Whole fruits) and vary your vegetables. Make half your grains whole grains, dairy should be low-fat and fat-free, eat different lean proteins.

Moderate physical activity should start out at minimum ten minutes per day and gradually work yourself up to one hour per day for best results. Things like walking, hiking, gardening, and yard work, stretching, and light weight training.
Then you can work up to more vigorous training such as jogging, running, swimming, aerobics, basketball, bicycling, and weight training.
Don't get obsessed with what you eat and working out, everything needs to be fun and in balance or you won't do it very long. Keep a food and activity journal. If you have a day where you eat too much or don't exercise, don't panic, this is for a lifetime.

My Self-esteem Calendar

In each block write down one thing that you like to do. Each block represents one day. Just one small thing that you would like to do for you each day. You can either draw a picture of the activity or write it down in the box that represents the day. For example, if you want to spend the day in the park with your dog, you can draw a picture of your dog in the block for the day that you would like to take your dog to the park or simply write it down in the block.

It's important that you make a commitment to yourself to check your self-esteem calendar every day to see what activity you have planned for yourself that day and follow through with it. You'll be glad that you did!

Confidence Building Exercise

1. _____
   Name

2. **<u>My Skills:</u>**

3. **<u>My Ambitions:</u>**

ACTION BREEDS CONFIDENCE & COURAGE.

4. **<u>What am I or will I be Famous for?</u>**

The Master Coaching Plan: The Core Competencies

1. The Foundation
- A. Based on Professionalism and Ethics.
- B. The Client and Coach Agreement.

2. The Relationship
- A. Establishing Trust & Intimacy between Coach and Client.
- B. Be the Alpha.

3. Communication
- A. Active Listening.
- B. Powerful Questions.
- C. Body Language, Tones, and Inflections. (Direct communication)

4. Discovery, Action, and Results
- A. Coach and Client Awareness.
- B. Action Design, Action Planning, and Goal Setting.
- C. Managing progress and Accountability Coaching.

The Initial Screening:
- 1. Usually you will be contacted by phone. On occasion, your contact will be through email, website, in person, or through a third-party referral.
- 2. The first thing you want to do is to get the person's name and contact information in case the call is dropped, so that you can get back in touch.
- 3. Establishing trust is important, so communicate honestly with the client. You need to have discernment and use good common sense depending on who you are talking with. If you're talking to the president of a Fortune 500 company you would not ask many of these questions over the phone. If you're talking to a guy that just got out of jail, of course you would.

4. There are some legal screening questions you will want to ask for your own protection:
- A. Client's Age?
- B. Client's Sex?
- C. Any Legal Trouble? If yes, what? Any Active Warrants?
- D. On Parole or Probation? If yes, Original Charges? In what State is the Parole or Probation?
- E. Any History of Violence? If yes, what?
- F. Any Mental Health Diagnosis? If yes, what is it?
- G. On any RX Medications? If yes, what are they?
- H. Any Physical Limitations? If yes, what are they?
- I. The Client's Address?

- ➢ J. Client's Occupation?
- ➢ K. How did they hear about you?
- ➢ L. Why Do They Want to Meet with A Life Coach?
- ➢ M. If the client's reason could put them at risk in any way, ask them if they are safe now? If No, take appropriate measure. Nine one one for Emergency!
- ➢ N. Do you feel that this client is a good match for you? Yes, or No.

Starting a coaching, client relationship:

1. The coach must clarify expectations with the client.
- ➢ A. Coaching Fees
- ➢ B. Coaching Schedule
- ➢ C. Coaching Agreement or Contract.
- ➢ D. Coaches and client's responsibilities
- ➢ E. What's Appropriate.
- ➢ F. What's Included in the Deal.

2. Establishing Trust with the client.
- ➢ A. The coach acts in a genuine, professional, and respectful manner.
- ➢ B. The coach demonstrates honesty and integrity.
- ➢ C. The coach follows all ethics and responsibilities.
- ➢ D. The coach is concise and keeps all commitments.
- ➢ E. The coach provides encouragement and support.
- ➢ F. The coach is sensitive to emotional areas of the client's life.

3. The coach must be able to be flexible, confident, and open in style.
- ➢ A. The coach must be able to go with the flow.
- ➢ B. The coach must be able to trust his or her own instincts.
- ➢ C. The coach must be a master of emotions.
- ➢ D. The coach must be willing to take risks.
- ➢ E. The coach must be able to see what works on the fly.
- ➢ F. The coach must be able to change the mood in the room.
- ➢ G. The coach must be able to shift perspectives.

4. Teamwork, Communication, and Consistency.
- ➢ A. The coach must be able to paint with words.
- ➢ B. The coach must use appropriate language, body language included.
- ➢ C. The coach must be able to explain exercises, purpose, techniques, etc.
- ➢ D. The coach must be able to reframe, so the client understands other perspectives.
- ➢ E. The coach must be able to provide clear feedback, summarizing, paraphrasing, and being a mirror for the client's words.

- ➢ F. The coach must focus on the client and what the client wants.
- ➢ G. Clarifies the client's goals, objectives, and beliefs as they stand.
- ➢ H. The coach must be a master of tones, inflections, and body language.
- ➢ The coach actively listens, building on the client's ideas.
- ➢ J. The coach is a master of powerful timely questions.
- ➢ K. The coach asks powerful questions that create clarity and open the mind.
- ➢ L. The coach asks questions that motivate fresh thinking, challenge outdated thinking, and assumptions.
- ➢ M. The coach generates a new energy and touches on a deeper vector through powerful questions.
- ➢ N. The coach uses open-ended questions, and linguistic architecture to help the client reach their objectives.

> 5. The coach must have the ability to accurately integrate and evaluate all information to assist the client to achieve the desired goals and objectives.

- ➢ A. Coach helps the client understand the difference between thoughts, feelings, emotions, and actions through interpretations and perceptions in regard to self and the world around self.
- ➢ B. The coach helps the client to discover new beliefs, perceptions, thoughts to help them achieve goals and objectives.
- ➢ C. The coach helps client by inspiring a shift in viewpoints.
- ➢ D. The coach brainstorms with the client to assist the client to define new actions that will deepen new learning and thinking.
- ➢ E. The coach challenges the client's perspectives on self and world view, and assumptions to help the client open their minds to new ways of thinking and solutions.
- ➢ F. The coach develops and helps to maintain effective coaching plan with the client.
- ➢ G. The coach works with the client to help them to set reasonable, attainable goals and objectives.
- ➢ H. The coach works with the client to establish a measurable, specific action plan with target dates and follow up.
- ➢ I. The coach helps the client identify and access different resources for learning to aid in success.
- ➢ J. The coach celebrates the wins with the client!
- ➢ K. The coach must continue to manage progress while leaving responsibility with the client.
- ➢ L. The coach recognizes growth and failures session to session, keeping client on track.
- ➢ M. Coach confronts client if they do not follow through as agreed.
- ➢ N. Coach helps client to develop ability to prioritize, address concerns, set pace, adjust as needed, get feedback.
- ➢ O. The coach responds to the client in the client's chosen medium of understanding, visual, verbal, etc.

Engagement:

- 1. The introduction where the coach and client share information about each other. The coach will clarify his training, experience, and certifications. The coach will clarify how coaching differs from therapy, mentoring, etc.
- 2. The client and coach develop their working agreement which establishes the parameters of the coaching process, fees, scheduling, the fact that sessions are client driven, duration, timing, and contract.

Action:

- 3. The coach sets the tone for the coaching work to be done. The environment should be conducive to learning, free from distractions, phones on vibrate, doors closed, ready to focus. If the coach is going to take notes during the session, the client should be made aware of this beforehand.
- 4. Some coaches will have the client figure out their goals and objectives in advance of the session; some do it at the first session. The two need to discover why the client felt they needed a coach, what does the client want to achieve?
- 5. Once the goals and objectives are identified, the two write an action plan together.
- A. What action will the client take to achieve their goals and objectives?
- B. When will the client start this action, and for what time period?
- C. How can the coach hold the client accountable?
- D. When can there be follow up on progress?
- E. How will the coach and client measure the client's progress?
- F. Who will give feedback on how the client is doing with the plan?
- G. Action Plans should be very specific with dates, times, people built in.

Managing Clint and Coach Progress and Accountability:

- 6. The coach will utilize active listening and powerful questioning, completely focusing on the client. The coach will utilize all his or her toolkit and skills to help the client see where they are stuck. Finding out the client's personal and world views, while building trust and respect with the client, summarizing, rephrasing, using open-ended questions, creating energy to move forward.
- 7. The coaching process must be results, process, and client driven.

Evaluating:

- 8. Evaluating each coaching session gives both coach and client a chance to reflect on the results, the relationship, and the process. They can see what is working and what is not and adjust appropriately.
- 9. Giving the client positive and behavioral feedback helps to build their strengths, thus making the coach more effective.

The Ending Stage: Disengagement:

- ➢ 10. Through skill and hard work we hope that the client breaks through and achieves their goals and objectives or at least gets close. If the sessions are running out, the coach may have to discuss additional sessions. The client may choose to end things as they are.
- ➢ 11. During the final session, the coach should point out the growth and change in the client. All victories should be celebrated along the way! The coach should make time to get feedback from the client and see if they would like to continue sessions. Don't continue just for money that would be a mistake, not to mention UNETHICAL!
- ➢ 12. If the coach did a good job, they will benefit in many ways, including word of mouth referrals. The coach should also do a self-evaluation to see where they can improve for their next client.

Establishing the Coaching Agreement

- Understanding what is required in the specific coaching interaction and then come to an agreement with the potential new client about the coaching process and relationship.
- The Coach fully explores client expectations from session establishing measurable success for the client in session and ensures clarity about coaching purpose.
- The Discussion: Specific guidelines and parameters, appropriateness, what's included, relationship defined, determines match between method and need.
- If all goes well and you decide to take on the client, send a client welcome letter that is warm, concise, and clear.
- The characteristics of the coaching relationship should be the same.
- Personal growth and change is affirmed, genuine support and encouragement, sincere concern, respect, shows interest and expresses belief in client goals. Coach trusts client to be responsible, client follows through on goals and action steps created.
- Coach believes the best about the client, nonjudgmental, gives room for client to fail.
- Client's strengths, uniqueness, and self-awareness are respected and encouraged.
- Exploration and Discovery are encouraged through active listening and powerful questioning. New possibilities and avenues.
- Trust, commitment, and integrity are crucial.
- Agreement content: What is coaching? Responsibilities, services, schedules, and fees, procedure.
- Confidentiality, release of information, cancellation, termination, limited liability.
- Coach's name and credentials, qualifications, certifications, etc., fees, date, etc.

The Art of Communication and Powerful Questioning as a Coach

The art of asking questions as it relates to professional coaching. A coach accompanies a client's dialog, where the client focuses on desired outcomes. For the coach, the desired goal is to create a space during the dialog within the client can grow. This is an emerging process. A dialog is when, through the words people speak, they structure the meaning. This is different than a discussion.

Coaches must consider that clients are experts in their own fields. In the coaching relationship, the client is thought to be the sole person capable of finding the answers to the set objectives. It really would be a fool's game to think that a client has not exhausted all ideas, thoughts, possible solutions, findings, and considered them, yet put them aside that any coach would think of or consider. Granted, this may be a little different when it comes to addictions coaching. In most cases, the coach will be the expert on recovery and the client the expert on using. In some cases, the client may be the expert on both they may have decades of experience with using and recovery.

Coaches need to be aware of traps and bad habits that will block them from success. Humility will go a long way in avoiding traps. Coaches need to put their egos in check and remember this is not about them, it's about the client. Clients aren't fools! The coach should not ask questions from the head, from the ego, or from pride. Coaches should not try to be tricky or try to outsmart clients.

Chances are that the client already tried to work things out on their own and failed. That's why they called a coach. An addict who already feels hopeless, will feel like they're at another dead end. They feel like there is no way to get sober. No way to change. They are totally hopeless! Their perception is where the problem sits. In their mind, their goals and objectives are not reachable.

That is why Professional Coaches do not focus on the problems as the client's way of describing the problem but do focus on the client's perspective of the problem. A well-defined problem will define its own solution, such as step one in the Basic Text of Alcoholics Anonymous gives us the problem, lack of power. If the problem is lack of power, the problem defines the solution, so the solution in step two must be power! A problem that can't find a solution has a "definition" problem.

The coach doesn't want to use the same process that the client did. The coach's job is to help the client to define the problem and/or the solution in a different way. Powerful coaching questions will accomplish this. Allowing the client to repeat all the details that reinforce their constraining frame of reference will reinforce the problem. The coach can get caught up in the problem thinking by continuing to listen and becoming in tune with client's emotions and feelings. The coach must question the client's frame of reference, and give them new original ones through powerful questioning.

Avoid questions starting with "and..." or "so..." these words may indicate linkage to preceding conversations. Does not interrupt the flow or help with new references. These are a red flag that the coach is getting too caught up in the client verbiage.

Keep things light and simple leaving room for the client to explore their frames of reference and have expanding inner dialog. Avoid leading questions; neutral open questions are much more effective. Avoid Negative Interrogations; ask positive-oriented coaching questions. Be aware of open and closed questions and when to use them. Closed questions should be used very carefully. Another red flag for coaches is rapid fire multiple questions without space in between for the client. Practical oriented questions can be very useful, esp. at the beginning of a coaching sequence. Active and analytic questions, analytic al questions generally elicit responses from the client's past. Coaches use these questions to help understand the client's motivations. Not favorable to help clients focus on future solutions and actions. Avoid "Why" questions: You will get a full accounting of the old frame of reference every time! "How" questions centered on action, considered good solutions oriented questions, unless prematurely asked. "When/Where" questions: ask where first then when falls into place. Impatient coaches ask when too early for selfish reasons.

Empower the client. Coaches need to remember to put the clients at the center of the coaching process. Focusing on achieving the client's goals and ambitions. Coaches accomplish this by formulating questions that suggest that clients take an active role in the relationship. Before coaches intrude into the private client space, professional coaches regularly gain respect by asking their clients for permission before asking closed personal questions.

1. Listening is the most powerful communication tool we have.

2. Body Language, Tones, and Inflections give us more information than the spoken word.
A. Words 7% of communication.

B. Tones and Inflections 38% of communication.

C. Body Language 55% of Communication

3. If a coach can master listening and body language they will be much more effective.

4. Asking Questions takes a great deal of knowledge and skill.

A. Ask open-ended questions.

B. Ask quality questions.

C. Questions should not come from your head.

5. Going Deeper

A. Question for the client: What is the most important thing in your life to resolve right now?

B. Question #2: What will happen if you don't?

6. We get a behavior and a belief.

A. Their limited beliefs don't serve them well.

B. Move them to a belief that is much more resourceful.

7. Natural law of cause and effect.

A. At effect, they are blaming.

B. At cause, they are taking responsibility.

C. If the answers outside of them, the problem is happening to them they believe.

D. They are powerless to change.

E. Question: Would you be open to the possibility that you haven't been handling the situation as well as you could have?

F. If they say "Yes!" Go deeper. (Now they can change!)

G. Question: How have you been handling the situation that hasn't been working?

H. Question: What do you think you could have done that would have been a much better solution to the situation?

I. Silence will lead to more thought and information from the client. When you both sit silently the client will become uncomfortable and feel that they are supposed to talk more. This is an interrogation technique called the Pause Technique.

J. You can change the meaning of a statement by the emphasis you put on certain words. For Example: A. I hate **liver**. B. **I** hate liver. C. I **hate** liver. If you put the emphasis on the bold word that becomes the focus of the statement. Same with a question Do **you** like work?

K. The more a coach works at his or her craft, the better they will get. Wisdom comes with time and experience. Practice your craft with peers, friends, and family. Role-playing is a great way to improve your skills.

Pay attention to body language, tones and inflections can give you extra information that will be very telling and assist you in your work. Although not an exact science, it is a pretty good indicator. For example, someone sitting with arms and legs crossed vs. someone sitting in an open position. The person that is crossed is defensive or cold! The clock watcher really isn't invested in the meeting. Someone that talks fast may be nervous, or laughter could be nervous or condescending, for seating chairs should be on a 45-degree angle. Usually when someone starts

55

to get real loud you have hit a nerve. Sometimes their voice volume will go very low because of guilt and shame. They may drop their head as an admission of guilt. Many factors come into play with reading body language such as culture, gender, age, etc.

looking right (generally)	eyes	creating, fabricating, guessing, lying, storytelling
looking left (generally)	eyes	recalling, remembering, retrieving 'facts'
bottom lip jutting out	mouth	upset
hand clamped over mouth	mouth / hands	suppression, holding back, shock
head tilted downward	head	criticism, admonishment
head tilted to one side	head	non-threatening, submissive, thoughtfulness
crossed (folded)	Arms or legs	defensiveness, reluctance
clenched fist(s)	hands	resistance, aggression, determination
uncrossed legs, sitting - general	legs or arms	openness

Some Examples of Body Language

The Art of Listening

Over time come experience and wisdom. A coach's skills will improve greatly if the coach works at his or her craft. When we get serious about listening to clients, we begin to hear different levels of information. Rookie coaches come up with the quick answer, the quick fix. A seasoned practitioner knows that is most often not the best answer. When the coach finds this understanding of communication, it allows the client to find a plan of action, an objective, and a solution.

Listening Rookie Level A

Our focus is on how the words the client is saying affect us. There is very little concern here for the client. The attention is on self. What are my thoughts? What are my issues? What are my judgments, emotions, feelings, conclusions, or findings? This is about collecting information, judgments and opinions.

Listening Experienced Level B

Our focus is much deeper on the speaker. We tune out everything else except for the speaker. We are totally aware of the person, what they say, how they say it, even what they don't say! We listen to what matters to the person, what makes them feel. We get out of our heads. When we ask a question it's a response to what we actually have heard.

Listening Master Level C

In addition to Levels A & B we listen even more deeply. We take everything into account. We use all of our senses. We learn to use are intuition. We use what's not being said, emotion, energy, all parties and form quality effective questions from all information including tones, inflections, body language, everything!

Then we articulate. Succinctly painting the same picture for the client that they painted for us. Describing what we learned from the client. Never judging, just clearly repeating what we believe the client means. This will make the client feel validated. "What I heard you say Cathy is this…"

If the client is not clear in their message, get clarity from them. Ask if you are not sure what they said or meant. The client will respect and like the fact that you are listening and want to be exact.

Be very curious about everything to go deeper. Never assume that you know the facts.

Remember the importance of silence, space, and simplicity. Use the pause technique when needed. Keep things simple and give the client the space to figure things out. To be an effective listener you must master this.

QUESTIONS:

- ➢ Are a prerequisite to knowledge, learning, and change.
- ➢ Take us into the future.
- ➢ Help us find solutions, goals, and objectives.
- ➢ Challenge old thinking and assumptions.
- ➢ Motivate change in thought.
- ➢ Show us insight and information.
- ➢ Get to the truth, past excuses and manipulations.
- ➢ Generate energy and deeper meaning.

Draw an imaginary line from the top to the base of the pyramid. The Top portion would have the highest power down to the ground which would be the lowest power.

The Coach's Power Pyramid!

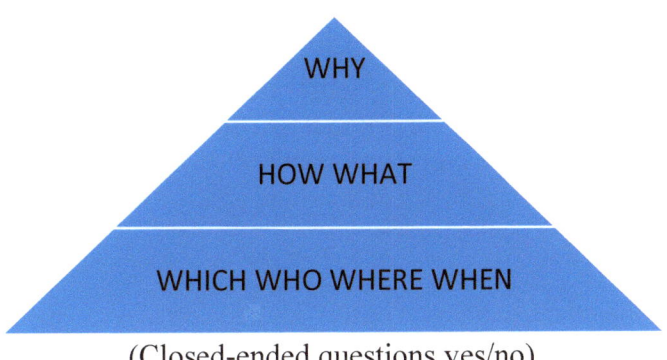

(Closed-ended questions yes/no)

"Why" is very Powerful but very Dangerous and selective when to use.

Think about how you could make the following question more powerful:

1. Are you going to become a life coach?
- A. Where are you training to become a life coach?
- B. How are you going to become a life coach?
- C. Why do you want to become a life coach?

Powerful Questions expand learning with a fresh perspective, action, and energy.

Some Examples of powerful questions:

- 1. If it had been you, what would you have done?
- 2. What have you tried so far?
- 3. What are your other options?
- 4. How does this fit in with your way of life?
- 5. How else could this have been handled?
- 6. If your life was depending on taking action, what would you do?
- 7. How can you improve the situation?
- 8. In the big picture, how important is this?
- 9. How does this relate to your life purpose?
- 10. What's occurred since our last session?

Questioning Technique:
- <u>Situation</u>: The Who, What, or Why of the experience or behavior.
- <u>Task</u>: Requires an explanation of what needs to be done or the task at hand.
- <u>Action</u>: How the task was accomplished or the situation overcome.
- <u>Result</u>: The person shares the associated outcome.

Paraphrasing Questions:

Shows speaker active and effective listening, understanding of the statement.

Should have three Elements:

- 1. Similar meaning or same thought as the original question.
- 2. It must elicit the same answer as the original answer.
- 3. Must show alternate wordings and order of some words.

> Take the time to allow the client to answer powerful questions. Why questions are good to solicit information. Don't use leading questions, do use open-ended questions. Stay away from Why questions they get people defensive and use How and What questions. Find out what the client knows about the problem. Ask them thought provoking questions. Pay attention, use intuition. Pay attention to body language, tones, and inflections.

- A. Words 7% of communication.
- B. Tones and Inflections 38% of communication.
- C. Body Language 55% of Communication

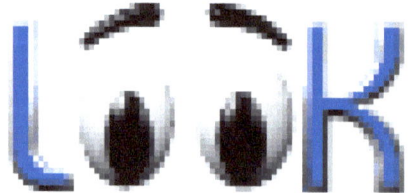

Legal Forms and Contracts for Professional Coaching

> *Please note all forms are for example only, and for the purpose of demonstration during training. You should consult an attorney before drafting any contracts or agreements on behalf of your company.

Cancelation Policy

If you fail to cancel a scheduled appointment, we cannot use this time for another client and you will be billed for the entire cost of your missed appointment.

A full session fee is charged for missed appointments or cancellations with less than a twenty-four-hour notice unless it is due to illness or an emergency. A bill will be mailed directly to all clients who do not show up for, or cancel an appointment.

Thank you for your consideration regarding this important matter.

_____ _____ ___/___/____
Print Client Name   Client Signature   Date

_____ _____ ___/___/____
Parent Name if under eighteen, Parent Signature if under eighteen   Date

_____ _____ ___/___/____
Professional's Name   Signature   Date

Coaching Agreement Contract

I_____ agree to the following contract with _____. I agree to a package of _____ one-hour sessions of Recovery Coaching at the fee of $_____.00 per hour. Each session can be scheduled in person, over the telephone, or on the computer such as Skype. I am responsible to be on time for appointments. Forty-eight hours notification to reschedule appointments in advance of scheduled appointments, otherwise the client will be charged in full for all scheduled appointments.

I agree to come to appointments without any illicit drugs or alcohol, weapons, or anything illegal in nature. I agree to come to appointments clean and sober. I agree to follow the directions of the assigned coach and complete all assignments on a timely basis. Payment is expected before each session. I understand that all coaches with _____ Coaches are Certified Professional Coaches.

Appointments

1._____
2._____
3._____
4._____
5._____

Signed: _____ Date: _____/_____/_____
Name: _____
Coach: _____ Date: _____/_____/_____

Confidentiality Agreement Contract

This Confidentiality Agreement (hereinafter the "Agreement") is made and entered into as of ,____/____/_____ by and between _____, with its principal place of business at _____(hereinafter the "Company"), and _____, with an address of _____(hereinafter the "Recipient"). The Company and the Recipient shall collectively be called the "Parties."

1. Purpose

The purpose of the Company disclosing the Confidential Information is to discuss progress…

Definition

"Confidential Information" means any information, technical data or know-how, including, but not limited to, that which relates to customers disclosed orally or in written or electronic form, and which is marked or identified by the disclosing party as "proprietary" or "confidential". Confidential Information does not include information, technical data or know-how which (i) is in the possession of the receiving party at the time of disclosure as shown by the receiving party's files and records immediately prior to the time of disclosure; or (ii) prior or after the time of disclosure becomes part of the public knowledge or literature, not as a result of any inaction or action of the receiving party, (iii) is approved for release by the disclosing party, or (iv) is independently developed by the receiving party without the use of any Confidential Information of the other party. Any information that is protected under the HIPAA regulations applies.

2. Nondisclosure of Confidential Information

The Company and Recipient each agree not to use the Confidential Information disclosed to it by the other party for its own use or for any unpermitted purpose. The recipient of Confidential Information will not disclose such Confidential Information to anyone, including to their employees; however, the recipient of Confidential Information may disclose such information to certain employees who are required to have such information to carry out the contemplated business. Each party has had, or will have employees, to whom Confidential Information of the other is disclosed, sign a Nondisclosure Agreement which is substantially similar to this Agreement and will notify the other in writing of the names of the persons who have had access to the Confidential Information of the other party. Each agrees that it will take all reasonable steps to protect the secrecy of and avoid disclosure or use of Confidential Information of the other in order to prevent it from falling into the public domain or the possession of unauthorized persons. Each agrees to notify the other in writing of any misuse or misappropriation of Confidential Information of the other that may come to its attention. Respecting all HIPAA regulations, requirements, and responsibilities.

Notwithstanding any other provision of the Agreement, disclosure of Confidential Information shall not be precluded if such disclosure:

(a) is in response to a valid order of a court or other governmental body of the United States or any political subdivision thereof;

(b) is otherwise required by law; or,

(c) is otherwise necessary to establish rights or enforce obligations under this Agreement, but only to the extent that any such disclosure is necessary.

(d) Involves Duty to warn, child abuse, elderly abuse, abuse of the mentally disabled.

(e) Involving suicidal ideation when the person expresses having a plan.

In the event that the receiving party is requested in any proceedings before a court or any other governmental body to disclose Confidential Information, it shall give the disclosing party prompt notice of such request so that the disclosing party may seek an appropriate protective order. If in the absence of a protective order, the receiving party is nonetheless compelled to disclose Confidential Information, the receiving party may disclose such information without liability hereunder; provided, however, that such party gives the disclosing party advance written notice of the information to be disclosed and upon the request and at the expense of the disclosing party, uses its best efforts to obtain assurances that confidential treatment will be accorded to such information.

3. Ownership

All Confidential Information shall remain the exclusive property of disclosing party, and recipient shall have no right to use Confidential Information except as provided herein. No patent, copyright, trademark, or other proprietary right or license is conveyed by this Agreement with respect to Confidential Information.

4. Return of Materials

Any materials or documents which have been furnished by one party to the other will be promptly returned, accompanied by all copies of such documentation, after the purpose of disclosure has been achieved. The receiving party further agrees to destroy all notes and copies thereof made by its officers and employees containing or based on any Confidential Information and to cause its agents and representatives to whom or which Confidential Information has been disclosed to destroy all notes and copies in their possession that contain Confidential Information upon the request of the disclosing party.

5. Intellectual Property Rights

Nothing in this Agreement is intended to grant any rights under any patent or copyright of either party, nor shall this Agreement grant either party any rights in or to the other party's Confidential Information, except the limited right to review such Confidential Information solely for the permitted purposes. The disclosing party warrants that it has the right to disclose its Confidential Information to the receiving party. Otherwise, all information is provided "as is" and without any warranty, express, implied or otherwise, regarding its accuracy or performance.

6. Independent Development

Each disclosing party understands that the receiving party may currently or in the future be developing information internally, or receiving information from other Parties that may be similar to the disclosing party's Confidential Information. Accordingly, nothing in this Agreement shall be construed as a representation or inference that the receiving party will not develop products, or have products developed for it, that compete with the products or systems contemplated by the disclosing party's Confidential Information.

7. Term

The confidentiality obligations of this Agreement shall remain in effect indefinitely and neither party will be allowed to disclose confidential information learned about the other at any point in the future unless required by law or court order.

8. Miscellaneous

This Agreement shall be binding upon and for the benefit of the undersigned Parties, their successors and assigns, provided that Confidential Information may not be assigned without consent of the disclosing party. This Agreement contains the final, complete, and exclusive agreement of the Parties relative to the subject matter hereof and supersedes any prior agreement of the Parties, whether written or oral. This Agreement may not be changed, modified, amended, or supplemented except by a written instrument signed by both Parties. Failure to enforce any provision of this Agreement shall not constitute a waiver of any term hereof.

9. Remedies

Each party agrees that its obligations hereunder are necessary and reasonable in order to protect the other party and the other party's business, and expressly agrees that monetary damages would be inadequate to compensate the other party for any breach of any covenant or agreement set forth herein. Accordingly, each party agrees and acknowledges that any such violation or threatened violation will cause irreparable injury to the other party and that, in addition to any other remedies that may be available, in law, at equity or otherwise, the other party shall be entitled to obtain injunctive relief against the threatened breach of the Agreement or the continuation of any such breach, without the necessity of proving actual damages. In such a case, the prevailing Party may request and be awarded attorney's fees and costs.

10. Notices

All notices hereunder shall be sent to either party at the address specified above, or such other address or contact person as the respective party may specify from time to time in accordance with the provisions of this Agreement.

In Witness Whereof, each of the Parties has signed this Confidentiality Agreement as of the date first above written.

Company

_____ ___/___/_____
       Print Name   Date

_____ ___/___/_____
       Signature   Date

Recipient

_____ ___/___/_____
       Print Name   Date

_____ ___/___/___
       Signature   Date

_____
Parent Print Name if Client Under eighteen

_____ ___/___/___
Parent Signature if Client Under eighteen   Date

Contract for Coaching Sessions

I_____ agree to the following contract with _____ I agree to a package of _____ one-hour sessions of Professional Coaching at the fee of $_____.00 per hour. Each session can be scheduled in person, over the telephone, or on the computer such as Skype. I am responsible to be on time for appointments. Forty-eight hours notification to reschedule appointments in advance of scheduled appointments, otherwise the client will be charged in full for all scheduled appointments.

I agree to come to appointments without any illicit drugs or alcohol, weapons, or anything illegal in nature. I agree to come to appointments clean and sober. I agree to follow the directions of the recovery coach and complete all assignments on a timely basis. Payment is expected before each session. I understand that all coaches with _____Coaches are Nationally Certified Recovery Coaches.

Appointments

1._____
2._____
3._____
4._____
5._____

Signed: _____ Date: _____/_____/_____
Name: _____
Coach: _____ Date: _____/_____/_____

Contract for Intervention Services

This is a contract entered into by _____, (hereinafter referred to as "the Provider") and _____ (hereinafter referred to as "the Client") on this date, ___/___/____. The Provider's location of services includes one intervention at the residence of the client, as dictated by the location and proximity of the client and provider.

The Client hereby engages the Provider to provide services described herein under "Scope and Manner of Services." The Provider hereby agrees to provide the Client with such services in exchange for consideration described herein under "Payment for Services Rendered."

Scope and Manner of Services

Services to Be Rendered by Provider: Addiction Intervention to include meeting with the family before client. Family session to include creation of impact letters, direction, staging, explanation of events and determination of who will attend Intervention. Direct Intervention to include client, family, and those appropriate for Intervention and agreed upon participations to include any therapists or psychiatrists, friends or significant others as deemed appropriate. Provider will provide drug and alcohol assessment for detox, and present information on residential recovery program. Provider will also provide transportation to the detox facility if needed or the residential facility. Time frame: Twenty-four hours from inception of family meeting.

Payment for Services Rendered

The Client shall pay the Provider for services rendered at $_____ for Intervention and services listed above. Due the day prior to the preintervention. Fees are for services and time spent regardless if client chooses to not attend treatment. All fees are nonrefundable.

Applicable Law

This contract shall be governed by the laws of the County of USA in the State of Pennsylvania and any applicable Federal law.

Signatures

In witness of their agreement to the terms above, the parties or their authorized agents hereby affix their signatures:

_____ (Family Rep.)   (Date) ___/___/___
_____ (Provider Rep.) (Date) ___/___/___

Contract for Intervention

This is a contract entered into by _____, (hereinafter referred to as "the Provider") and _____ (hereinafter referred to as "the Client") on this date, ____/____/_____.

The Provider's location of services includes one intervention at the residence of the client, as dictated by the location and proximity of the client and provider.

The Client hereby engages the Provider to provide services described herein under "Scope and Manner of Services." The Provider hereby agrees to provide the Client with such services in exchange for consideration described herein under "Payment for Services Rendered."

Scope and Manner of Services:

Services to Be Rendered by Provider: Addiction Intervention to include meeting with the family before client. Family session to include creation of impact letters, direction, staging, explanation of events and determination of who will attend Intervention. Direct Intervention to include client, family, and those appropriate for Intervention and agreed upon participations to include any therapists or psychiatrists, friends or significant others as deemed appropriate. Provider will provide drug and alcohol assessment for detox, three choices of rehabs and Intervention to present to client finding to attend detox or rehab. Provider will also provide transportation to rehab including flight or car ride (tailor to fit situation).

Time frame: Twenty-four hours from inception of family meeting.

Payment for Services Rendered:

The Client shall pay the Provider for services rendered at $_____.00, the amount of _____ dollars and zero cents for Intervention and services listed above. Half due along with transport fees of $_____.00, _____ dollars and zero cents to include $_____.00, _____dollars and zero cents upfront and remaining balance due at time of family meeting. Fees are for services and time spent regardless if client chooses to not attend treatment. All fees are nonrefundable.

Applicable Law:

This contract shall be governed by the laws of the County of USA in the State of _____ and any applicable Federal law.

In witness of their agreement to the terms above, the parties or their authorized agents hereby affix their signatures:

_____ _____ ___/___/_____
Client Print Name   Signature   Date

_____ _____ ___/___/_____
Provider/ Interventionist   Signature   Date

Note if client is under eighteen years of age, a legal guardian or parent must sign for them.

Any special instructions outside of contract:
_____
_____
_____.

Random Drug Test Sheet

Date of Test: ___/___/____

Tester: _____
Certifications: _____

Client Name: _____ ___/___/____
                    Print Name   Date

              _____ ___/___/____
                    Signature   Date

Results: Negative ☐ Positive ☐

If Positive for what substance? COC ☐ THC ☐ MET ☐ OPI ☐ AMP ☐

Tester: _____ Date: ___/___/____
            Print Name
      _____ Date: ___/___/____
           Signature Name

Incident Report: _____
_____
_____
_____
_____
_____
_____
_____
_____.

_____ _____ ___/___/____
Staff Print Name   Staff Signature   Date

Drug Testing Consent Form

I, _____, have read and understand the substance abuse and drug testing policy of _____ and hereby agree to comply with them, the procedure involved, and do hereby freely and voluntarily give my consent to the testing laboratory designated by _____ to perform analytical tests deemed necessary to determine the absence or the presence of alcohol and drugs in my system. I understand that if the sample I provide does test positive, I will be subject to denial of employment or termination of employment, to the extent permitted by applicable law.

I understand that _____ may request proof that I am taking a controlled substance as directed pursuant to a lawful prescription issued in my name. If requested, I must provide such proof within days.

I understand that the testing to be performed will include, but may not be limited to the following: Alcohol, Amphetamines, Barbiturates, Cocaine, Opiates, Phencyclidine, Marijuana, Benzodiazepines, Methadone, Propoxyphene, Urine Ethanol, Oxycodone, and Meperidine and such results will promptly be submitted to _____.

To avoid false positive results, I am advised to avoid foods with poppy seeds within seventy-two hours prior to testing. I am advised to refrain from drinking more than forty ounces of liquid within three hours prior to providing the urine sample for testing. I understand the recommendations listed above, and I am aware that a sample reported as "diluted," constitutes a positive result. A positive result will be confirmed as follows (or as otherwise required by applicable law):

A positive finding of drugs by preliminary screening procedures will be subsequently confirmed by gas chromatography-mass spectrometry or other scientific testing technique which has been or may be approved by the appropriate state or other governmental office.

Furthermore, I understand that in the event a sample is reported as "diluted," I may retest within three days at my own cost. If the resulting sample tests negative, I understand that _____ will reimburse me for the cost of the test.

I realize that the results of this testing will be shared to the extent required or permitted by applicable law, which disclosure may include, but is not limited to, me, and such officers, agents or employees of _____ who need to know the information for reasons connected with their employment.

My signature below signifies my consent to provide urine samples for testing to determine the presence of drugs in my body, as well as my consent to have the results of such results disclosed as described above, and my release of _____, its directors, officers, employees, agents, successors and assigns, and my agreement to hold each of them harmless from, any and all liabilities, claims, losses, costs and expenses whatsoever (including attorney's fees and court costs) arising out of or resulting from any such testing or disclosure.

_____	_____	___/___/___
Name	Signature	Date
_____	_____	___/___/___
Witness Name	Signature	Date

Drug Testing Consent Form

I, _____, have read and understand the substance abuse and drug testing policy of _____ and hereby agree to comply with them, the procedure involved, and do hereby freely and voluntarily give my consent to the testing laboratory designated by _____ to perform analytical tests deemed necessary to determine the absence or the presence of alcohol and drugs in my system. I understand that if the sample I provide does test positive, I will be subject to denial of employment or termination of employment, to the extent permitted by applicable law.

I understand that _____ may request proof that I am taking a controlled substance as directed pursuant to a lawful prescription issued in my name. If requested, I must provide such proof within days.

I understand that the testing to be performed will include, but may not be limited to the following: Alcohol, Amphetamines, Barbiturates, Cocaine, Opiates, Phencyclidine, Marijuana, Benzodiazepines, Methadone, Propoxyphene, Urine Ethanol, Oxycodone, and Meperidine and such results will promptly be submitted to _____.

To avoid false positive results, I am advised to avoid foods with poppy seeds within seventy-two hours prior to testing. I am advised to refrain from drinking more than forty ounces of liquid within three hours prior to providing the urine sample for testing. I understand the recommendations listed above, and I am aware that a sample reported as "diluted," constitutes a positive result. A positive result will be confirmed as follows (or as otherwise required by applicable law).

A positive finding of drugs by preliminary screening procedures will be subsequently confirmed by gas chromatography-mass spectrometry or other scientific testing technique which has been or may be approved by the appropriate state or other governmental office.

Furthermore, I understand that in the event a sample is reported as "diluted," I may re-test within three days at my own cost. If the resulting sample tests negative, I understand that _____ will reimburse me for the cost of the test.

I realize that the results of this testing will be shared to the extent required or permitted by applicable law, which disclosure may include, but is not limited to, me, and such officers, agents or employees of _____ who need to know the information for reasons connected with their employment.

My signature below signifies my consent to provide urine samples for testing to determine the presence of drugs in my body, as well as my consent to have the results of such results disclosed as described above, and my release of _____, its directors, officers, employees, agents, successors and assigns, and my agreement to hold each of them harmless from, any and all liabilities, claims, losses, costs and expenses whatsoever (including attorney's fees and court costs) arising out of or resulting from any such testing or disclosure.

_____	_____	___/___/_____
Name	Signature	Date
_____	_____	___/___/_____
Witness Name	Signature	Date

Informed Consent

I have chosen the services of _____ Coaching to provide _____ Coaching services, Case Management, Intervention Services, Sober Companion services, or _____ services. I understand that for all legal and regulatory purposes, the services I am provided will be considered to be provided in the State of _____ and the United States.

I hereby certify that I am of legal age of consent according to the laws of my resident state, province, or country. Which means I am eighteen years of age or older. I understand confidentiality is importance to both the Professional and the client. I will keep confident all sessions regardless of the mode of services provided, whether they be by phone, in person, or over the internet.

A. All Professionals from _____ will keep all issues discussed confidential. I understand that the only appropriate breach of confidentiality is when my Professional _____ believes that I may intend to harm or seriously injure myself or other individuals, and items included in section B, C, and D.

B. Should I be involved in child abuse, child neglect, spouse abuse or elder abuse, or abuse of a physically or mentally disabled person. I also understand that if I display suicidal ideations and a plan.

C. Any communication over the internet or telephone cannot be guaranteed to be HIPAA protected.

D. I understand that I cannot come to appointments with anything illegal including weapon, illicit drugs, contraband of any kind, liquor, and that I must not be intoxicated, threatening, violent, or disrespectful. If I have any active warrants I am to immediately make that known.

_____   _____   ___/___/___
Printed Name   Signature   Date

_____   _____   ___/___/___
Professional's Printed Name   Signature   Date

Medical Record Release Form

I, _____, born on, ___/___/_____ hereby authorize, _____ to release to, _____ the medical information from my personal medical records listed below.

Medical Records Released:

I authorize to release to the following medical records:
_____
_____
_____
_____

Purpose of Release:

I give my permission for this medical information to be used for the following purpose:
_____
_____
_____

I do not give permission for any other use or re-disclosure of this information.

Permission Expires on ___/___/_____

_____   _____/_____/_____
Patient Signature   Date

_____   _____/_____/_____
           Witness   Date

Medical Release Form

I, _____, born on ___/___/_____, Social Security Number _____-____-_____, hereby authorize Dr. _____ to release to _____, the medical information from my personal medical records listed below.

Medical Records Released:

I authorize Dr. _____ to release to _____ the following medical records:

1. Information on all medications and reason for use.
2. Medical clearance to live in a residential recovery program.
3. Medical clearance to be involved in nonclinical addiction recovery, case management, recovery coaching, etc.

Purpose of Release:

I give my permission for this medical information to be used for the following purpose:

Fitness evaluation and items one through three listed above.

I do not give permission for any other use or re-disclosure of this information.

Expiration Date:

This Medical Release Authorization will expire on ____/____/_____

Future Medical Records:

I give permission for the following medical records that will be created in the future to be released to _____:

Any Rx medications and reason.

_____   ____/____/_____
Client Name   Date

_____
Client Signature

Medical Record Release Form

I, _____ , born on, ___/___/_____ hereby authorize, _____ to release to, _____ the medical information from my personal medical records listed below.

Medical Records Released:

I authorize to release to the following medical records:

_____
_____
_____
_____

Purpose of Release:

I give my permission for this medical information to be used for the following purpose:

_____
_____
_____

I do not give permission for any other use or re-disclosure of this information.

_____ _____/_____/_____
Patient Signature Date

_____ _____/_____/_____
Witness Signature Date

Medical Record Release Form

I, _____ , born on, ___/___/_____ hereby authorize, _____ to release to, _____ the medical information from my personal medical records listed below.

Medical Records Released:

I authorize to release to the following medical records:

_____
_____
_____
_____

Purpose of Release:

I give my permission for this medical information to be used for the following purpose:

_____
_____
_____

I do not give permission for any other use or re-disclosure of this information.

_____   _____/_____/_____
Patient Signature   Date

_____   _____/_____/_____
Witness Signature   Date

Minors Guardianship

Clients that are minors, under the age of eighteen and non-emancipated are subject to current laws as to limits of confidentiality in regard to parents and legal guardians.

Parents or legal guardians of non-emancipated minor clients have the right to access the clients' records.

I agree to the above limits of confidentiality and understand their meanings and ramifications.

_____ _____ _____ ___/___/___
Client Name   Age   Signature   Date

_____ ___/___/___
Client's Parent/Guardian if under eighteen   Date

Payment Agreement

This payment agreement (the "Agreement") is made on ___/___/____ by and between:

_____ residing at _____,
_____, _____, hereinafter referred to as the "Client"; and

_____ residing at _____,
_____, _____, hereinafter referred to as the "Provider", and together with the Client the "Parties".

In consideration of services purchased under the terms and conditions of this Agreement and other good and valuable consideration, the Parties agree as follows:

On the execution of this Agreement the Provider agrees to Provide to the Client services valued in the dollar amount of $_____.00 (the "Agreed Fee") on the terms and conditions set out in this Agreement. Services are to include: _____
_____.

If payment is being provided for a third party, the services will be provided to the following: _____ residing at _____
_____.

The fee of _____ dollars are due on the _____ day of service provided and have been calculated to allow the Agreed Fee to be paid in full on an Hourly/ Daily/ Weekly/ or Monthly Basis within ____ Days/ Weeks/ or Months or end of service.

The Client may pay the Provider within three days after the date of this Agreement without penalty. (Weekly or Monthly Service Only)

This Agreement shall be governed in accordance with the laws of the State of _____, applicable to agreements to be wholly performed therein.

In Witness hereof these presents are executed on the date before written:

_____ ____/____/_____
Client

_____   Date

Client if under eighteen Parent

_____ ____/____/_____
Provider   Date

_____ ____/____/_____
Signed in the presence of: Date
Witness

_____ ____/____/_____
Name: Date

Address:_____,_____
,_____,_____.

Coaching Services Agreement (Example)

I. This Coaching Services Agreement (the "Agreement") is entered into as of by and between The Coaching Company (the "Contractor"), an individual, and John Doe (the "Hiring Party", sometimes hereinafter collectively referred to as the "Parties").

II. Whereas, the Hiring Party and Contractor hereby enter into this Agreement whereby Contractor will render certain services to and for the benefit of the Hiring Party in exchange for valuable consideration.

III. Now, therefore, for and in consideration of the mutual covenants contained herein and other good and valuable consideration, the receipt and sufficiency of which is hereby acknowledged, the Hiring Party and Contractor do hereby contract, covenant and agree as follows:

    A.   **Agreement.** Contractor does hereby agree to render and provide services in accordance with the terms of this agreement and as specified in Paragraph "E" herein.

    B.   **RATE.** The Hiring Party does hereby agree to pay Contractor the rate of $250.00 per hour.

    C.   **Independent Contractor.** Contractor is, and will continue to be for the duration of this Agreement, an independent contractor and is not to be considered in any way subject to control by the Hiring Party. Contractor is not, and is not to be considered, an agent or employee of the Hiring Party.

    D.   **Indemnity.** Contractor does hereby for himself/herself, and his/her heirs, executors, administrators, officers, employees, subcontractors, successors and assigns, agree and covenant to indemnify, save and hold harmless the Hiring Party and his or her heirs, executors, administrators, agents, employees, attorneys, successors and assigns from any and all claims, demands, actions, causes of action, suits at law or in equity, damages, costs, expenses, and losses of any kind or nature whatsoever, whether now known or unknown, which may not exist or which may hereafter arise out of or from the work, services, labor and/or materials to be rendered and provided by Contractor or its subcontractors to or for the benefit of the Hiring Party.

    E.   **Description of Service to Be Performed.** Contractor agrees to perform and/or provide the following services to the Hiring Party: Services provided will be scheduled at least 7 days in advance of any appointment with the client. Twenty-four-hour notice to cancel an appointment, otherwise the client is expected to pay for as scheduled appointment. If the client is late, they will be given a -

minute grace period before being charged. The services provided will be coaching services to assist the client to achieve desired goals and objectives through the use of recognized core competencies, tools, and skills of the professional coaching industry.

F. **Agreement Term.** This Agreement shall commence on and shall continue Six Months. This Agreement may be terminated by the Parties as follows: May be terminated by either party upon giving thirty days written notice. In the event that either party breaches any term of this Agreement, such breach shall operate to terminate this Agreement as between the Parties, and the non-breaching party shall have no further obligations there under. However, all unperformed obligations of the breaching party will remain due and owing.

G. **Confidentiality.** In the course of performing the services as described herein, the Parties acknowledge that the Contractor may come in contact or become familiar with information which the Hiring Party may consider private, proprietary and confidential. Contractor agrees to keep all such information confidential and not to discuss or divulge it to anyone other than appropriate Hiring Party family members, personnel or their designees.

H. **Contractor's Taxpayer I.D. Number.** The taxpayer I.D. number of the Contractor is 111-22-3333. If applicable, necessary or required, the Contractor is licensed to perform the agreed upon services enumerated herein and covenants that he or she maintains all valid licenses, permits and registrations to perform the same.

I. **Competent Performance of Services.** Contractor agrees that all services will be done in a competent fashion in accordance with applicable standards of the Contractor's profession or trade and all services are subject to final approval by a representative of the Hiring Party prior to payment.

J. **Representations and Warranties.** The Contractor will make no representations, warranties, or commitments binding the Hiring Party without the Hiring Party's prior written consent.

K. **Legal Right.** Contractor covenants and warrants that he/she has the unlimited legal right to enter into this Agreement and to perform in accordance with its terms without violating the rights of others or any applicable law and that he/she has not and shall not become a party to any other agreement of any kind which conflicts with this Agreement. Contractor shall indemnify and hold harmless the Hiring Party from any and all damages, claims and expenses arising out of or resulting from any claim alleging that this Agreement violates any such other agreements. Breach of this warranty shall operate to terminate this

Agreement automatically without notice and to terminate all obligations of the Hiring Party to pay any amounts which remain unpaid under this Agreement.

L.   **Waiver.** Failure to invoke any right, condition, or covenant in this Agreement by either party shall not be deemed to imply or constitute a waiver of any other rights, conditions, or covenants, and neither party may rely on such failure.

M.   **Additional Terms.**

    a. **Entire Agreement and Amendments.** This Agreement constitutes the entire agreement of the parties with regard to the subject matter hereof, and replaces and supersedes all other agreements or understandings, whether written or oral. No amendment or extension of this Agreement shall be binding unless in writing and signed by both parties.

    b. **Binding Effect, Assignment.** This Agreement shall be binding upon and shall inure to the benefit of Contractor and the Hiring Party and to the Hiring Party's successors and assigns. Nothing in this Agreement shall be construed to permit the assignment by Contractor of any of its rights or obligations hereunder, and such assignment is expressly prohibited without the prior written consent of the Hiring Party.

    c. **Governing Law, Severability, Attorneys' Fees.** This Agreement shall be governed by the laws of the State of NY. The invalidity or unenforceability of any provision of this Agreement shall not affect the validity or enforceability of any other provision. In the event that a dispute arises involving the subject matter of this Agreement, the prevailing party in such dispute shall be entitled to their reasonable attorneys' fees and costs.

Wherefore, the parties have executed this Agreement as of the date stated above.

By:

_____

John Doe
111 Spirit Drive
Miami, FL11111

By:

_____

The Coaching Company
222 Solution Way
New York, NY33333

Professional Coaching Contract

I_____ agree to the following contract with _____ Coaching. I agree to a package of _____ one-hour sessions of Recovery Coaching at the fee of $_____.00 per hour. Each session can be scheduled in person, over the telephone, or on the computer such as Skype. I am responsible to be on time for appointments. Forty-eight hours notification to reschedule appointments in advance of scheduled appointments, otherwise the client will be charged in full for all scheduled appointments.

I agree to come to appointments without any illicit drugs or alcohol, weapons, or anything illegal in nature. I agree to come to appointments clean and sober. I agree to follow the directions of the recovery coach and complete all assignments on a timely basis. Payment is expected before each session. I understand that all coaches with _____Coaches are Nationally Certified Recovery Coaches.

Appointments:

1._____ TIME _____AM/PM
2._____ TIME _____AM/PM
3._____ TIME _____AM/PM
4. _____ TIME _____AM/PM
5._____ TIME _____AM/PM

Parent if under eighteen _____ Date ____/____/_____
Signed_____ Date ____/____/_____
Name_____
Coach _____ Date ____/____/_____

Professional Services Agreement

This Professional Services Agreement (the "Agreement"), dated,___/___/____ between_____ of _____ ,_____, (the "Professional") and _____ of_____ ,_____, (the "Client").

Whereas the PROFESSIONAL is engaged in the business of providing professional services in the field of_____in the State of PA and elsewhere; and

Whereas the Client desires to avail itself of these professional services of the Professional from time to time in connection with the Client's business activities and the Professional desires to enter into this Agreement with the Client.

Now therefore in consideration of the mutual promises and agreements contained in this Agreement, and other good and valuable consideration, the parties agree as follows:

1. **Object:** The Professional shall furnish to the Client his, her or its professional services in accordance with the details and specifications as identified on Schedule "A" attached hereto (such Schedule constituting integral terms of the Agreement). The Professional shall perform such professional services at all times in accordance with the commonly accepted standards of the Professional's profession, trade or craft and in full compliance with the statutes, laws, ordinances and regulations governing the Professional's profession, trade, craft or business from a work location situated in PA.

2. **Independent Professional:** The Professional shall have the sole authority to dictate direction of the work covered by this Agreement and shall be responsible for the manner in which the said work is done, for the method employed in doing the same and for all acts and things done in the performance of the Professional's obligations hereunder, except for departing from the Professional's normal practices which may be requested by the Client from time to time. Nothing contained in this Agreement and the relationship created between the parties hereby shall, directly, or indirectly, constitute the Professional as an agent or an employee of the Client and further, nothing herein shall operate or be construed to relieve the Professional of any duties or obligations imposed upon it as an independent professional.

3. **Taxes:** The Professional shall be responsible to withhold or deduct premiums, taxes, or levies as the case may be as required under Federal and State law and the Professional shall be responsible to withhold and remit any deductions for taxes, levies or contributions imposed by any authority with respect to both the remuneration paid under this Agreement and the work incidental thereto.

4. **Professional's Fee:** The Client shall pay the Professional for his, her, or its services a fee of _____ Dollars (the "Fee"), payable within____days of Client's receipt of Professional's invoice. All fee paid are nonrefundable for any reason.

5. **Term:** This Agreement shall be deemed to have come into force and effect on ___/___/____ and shall continue through ___/___/____ (the "Term"). These dates may be delayed upon the written consent of both parties. Nothing in this paragraph shall be construed as affecting the rights of the parties to terminate this Agreement at an earlier date in accordance with sections in this Agreement pertaining to termination.

6. **Termination For Cause**

   A. If either party to this Agreement is in breach of any of its obligations under this Agreement, the other party may give a notice in writing of the breach to the defaulting party and request the latter to remedy it. If the party in breach fails to remedy the breach within thirty days after the date of written notice, then this Agreement may be terminated immediately by written notice of termination given by the complaining party.

   B. The Client may terminate this Agreement by written notice to take effect immediately upon receipt of the notice by the Professional if:

   (i) the Professional is in breach of the provisions contained herein relating to the secrecy of confidential matters of this Agreement; or

   (ii) the Professional becomes insolvent or bankrupt or makes an assignment for the benefit of creditors, or a receiver is appointed of its business; or a voluntary or involuntary petition in bankruptcy is filed or proceeding for the reorganization or winding-up of the Professional's business is instituted; or

   (iii) the Professional attempts to assign or cede an interest in this Agreement without the prior consent of the Client; or

   (iv) if the Professional comes under the direct or indirect control of any company or person who does not control it at the date of execution of this Agreement; or

   (v) if the Professional is grossly negligent in carrying out its duties hereunder; or

   (vi) if the Professional becomes unable to discharge his or her duties by reason of mental or physical illness or disease for a period of two consecutive months or more, or should he or she become permanently disabled and unable to fulfill his or her duties and the Professional does not find a replacement professional who is completely satisfactory to the Client in its sole discretion; or

   (vii) if the Professional or the Professional's employees are engaged in any fraudulent or illegal activity.

C. The provisions of this section shall not in any way restrict the rights of either party hereto to terminate this Agreement pursuant to any other paragraph in this Agreement.

7. **Early Termination Without Cause**

    A. The Client may terminate this Agreement for any reason by giving thirty days' written notice of the Client's intent to terminate. For termination to be effective the Client must pay, prior to the date of termination, all outstanding invoices, fees, and reimbursements, if applicable, due for payment. Additionally, if applicable, Client must pay, prior to the date of termination, a prorated share of all fees due and owing either based upon time spent if the payment of fees is made on an hourly basis or work completed if the payment structure is a flat-fee agreement.

    B. The Professional may terminate this Agreement for any reason by giving thirty days' written notice of the Professional's intent to terminate. For termination to be effective the Professional must provide to the Client twenty days prior to the date of termination all outstanding invoices and requests for reimbursements, if applicable, then due for payment and a schedule of work that will be performed and completed by the termination date. Professional agrees, if the Professional initiates termination of the Agreement pursuant to this paragraph, that all unpaid fees due on the date of termination will be paid on a prorated basis calculated based upon time spent if the payment of fees is made on an hourly basis or work completed if the payment structure is a flat-fee agreement.

8. **Assignment:** It is expressly agreed that this Agreement shall not be assigned or transferred, in whole or in part, by either of the parties hereto without the prior express written consent of the other Party.

9. **Confidential Information**

    A. The Professional acknowledges that during the performance of the services described herein the Professional may come in contact with, be exposed to or come into possession with, information that is deemed to be confidential, private, proprietary or otherwise secret to the Client and that is not general known to the general public. The Professional expressly agrees that the Professional will keep any such information confidential and will not divulge such information to any persons or entities unless the Client gives its express written permission to do so and such divulging of information is necessary to perform the services that are the subject of this Agreement. Further, except as may be necessary in the performance of the services described within this Agreement, the Professional shall not at any time or in any manner make or cause to be made any copies, pictures, duplicates, facsimiles or other reproductions or recordings of any type, or any abstracts or summaries of any reports, studies, memoranda, correspondence, manuals, records, plans or other written, printed or otherwise recorded material of the Client, or which relate in any manner to the present or prospective business of the Client. The Professional shall have no interest in any of this material and agrees to surrender any and all of the material which may be in its possession to the Client immediately upon the request of the Client.

    B. The Professional shall not at any time except under legal process divulge any matters relating to the business of the Client or any customers or agents of the Client which may become known to it by reason of its services hereunder and shall be true to the Client in all dealings and transactions relating to the services contemplated by this Agreement. Furthermore, the Professional shall not use at any time (whether during the continuance of this Agreement or after its termination) for its own benefit or purposes or for the benefit or purposes of any other person, firm, corporation, association or other business entity, any trade secrets, business development programs, or plans belonging to or relating to the affairs of the Client, including knowledge relating to customers, clients, or employees of Clients.

10. Notices: Wherever in this Agreement it shall be required or permitted that notice be given or served by either party to or on the other, the notice shall be in writing and shall be delivered personally to the party to whom it is given or sent by prepaid, registered mail, addressed as follows:

To the Professional at:

,
Email:
Facsimile:

to the Client at:

,
Email:
Facsimile:

And each such notice shall be deemed given three business days after mailing in the case of mail and two hours after the time of transmission in the case of facsimile or email transmission. This address and/or facsimile numbers or email addresses may be changed from time to time by either party by notice as above provided.

11. Interpretation

    A. This Agreement constitutes all of the agreements between the Professional and Client pertaining to the subject matter of it and supersedes all prior agreements, undertakings, negotiations and discussions, whether oral or written, of the parties to it and there are no warranties, representations or other agreements between the parties to it in connection with the subject-matter of it except as specifically set forth or referred to in this Agreement. No supplementation, modification, waiver, or termination of this Agreement shall be binding unless executed in writing by the party hereto to be bound thereby. No waiver of any other provisions of this Agreement shall be deemed or shall constitute a continuing waiver unless expressly provided.

B. Headings are not to be considered part of this Agreement, and are included solely for convenience of reference and are not intended to be full or accurate descriptions of the contents of any section.

C. In this Agreement, words importing the singular number include the plural and vice versa, words importing the masculine gender include the feminine and neuter genders; and words importing persons include individuals, and proprietors, corporations, partnerships, trusts, and unincorporated associations.

D. This Agreement shall be governed by and construed in accordance with the laws of the State of PA.

E. The invalidity or unenforceability of any provision of this Agreement or any covenant in it shall not affect the validity or enforceability of any other provision or covenant in it and the invalid provision or covenant shall be deemed to be severed.

Time Being of the Essence

A. Time shall be deemed to be of the essence of this Agreement; provided from time to time for completing any work, which has been or is likely to be delayed by reason of *force major* or other cause beyond the reasonable control of the Professional, shall be extended by a period equal to the length of the delay so caused, provided that prompt notice in writing of the occurrence causing of likely to cause such delay is given to the Client.

B. The Client shall advise the Professional in writing of any occurrence causing or likely to cause delays in the completion of its responsibilities under this Agreement.

12. **Title to Work Being Performed:** Upon payment being made in accordance with the terms of this Agreement, all title, rights and interest in all printed materials and other physical media, containing designs, symbols, inventions and reports performed, created or written in accordance with this Agreement shall vest in and ensure to the benefit of the Client, it being understood that such vesting of title shall not constitute acceptance by the Client of such work in conformity with the specification or requirements of the Agreement. Without restricting the generality of the foregoing, the right of publication of any research paper or study performed under this Agreement shall vest solely in the Client upon payment as aforesaid, and any person desiring to publish any such research or study, in whole or in part, shall first obtain the written permission of the Client.

13. **General:** This Agreement shall ensure to the benefit of and be binding on the parties hereto and their respective heirs, executors, administrators, successors and assigns.

In Witness Whereof the parties have hereunto set their respective hands and seals as at the date written above.

                                                Professional:
                                                Per:

                                                _____

                                                Name:
                                                Title:
                                                I have authority to bind the Company.

                                                Client:
                                                Per:

                                                _____

                                                Name:
                                                Title:
                                                I have authority to bind the Company

Schedule "A"

Specified Services of the Professional
(Schedule "A" stands by itself as a report of services rendered by the professional or the professional's company.)

Sober Coach/ Sober Companion Live-In Agreement

I, _____ make a commitment of 1, 2, 3 months to participate in the _____ Sober Coaching/ Companion Program. *(circle one)*

I agree to pay $_____.00 per month each month for the terms of my commitment. Agreed amount written: _____ dollars and zero cents). This amount is due by the first of each month. All payments must be on time! There will be a $50.00-dollar late fee for any payments arriving more than two days after the due date.

Signed: _____ Date: ____/____/____ Witness: _____

I understand that I am responsible to report any emergencies or issues immediately to the staff. I understand there is zero tolerance for violence, weapons, illicit drugs or alcohol, discrimination, abuse, drug or alcohol use, smoking indoors, theft, non-payment. I understand that I must continue to attend meetings and work with a coach from the program, and keep all coaching appointments. I understand that I must submit to drug and alcohol screening. I understand that all deposits and payments are non-refundable. I understand that I am to vacate the premises immediately if deemed necessary by a representative of _____ for any of the above listed reasons and return to residential treatment. No person shall sign this agreement who knowingly has an active warrant from any judicial system.

I hereby agree to the above listed contract: Date: ____/____/____

Name: _____   Witness: _____

Signature: _____   Witness: _____

S.S. #: _____-_____-_____   Cell #:(____) _____-_____

DOB: ____/____/_____   Emergency #: (____) _____-_____

Coach: _____   Date: ____/____/____

Financial Agreement

I, _____ agree to pay JKY Corporation the sum of $_____.00

Amount of payment in written form: _____dollars and zero cents for the following services:
_____.

I, _____ agree to make an initial non-refundable deposit by the following date: ___/___/____. I also agree to pay the full balance owed by the following date: ___/___/____. I understand and agree that all payments are non-refundable. The services I am paying for are for

Myself, or the following person: _____. I take full financial responsibility for any damages caused at the above listed facility by myself, or by the client listed in this agreement, or any other expenses incurred by the facility, such as transportation to any external appointments, airport, or bus pick up, or any other cost to the facility. Should there be an agreement to extent the initial stay, I agree to make all payments of fees on a timely basis. Should I not make all payments on a timely basis, I agree to pay all lawyer fees and court costs incurred by JKY Corporation to settle the matter. Service is to begin on the following date: ___/___/____ and end on ___/___/____.

Should the initial stay be extended the new exit date will be ___/___/____.

_____ ___/___/____ _____ ___/___/____
Client Signature   Date   Witness Signature   Date

___-___-____ ___/___/____ _____ ___/___/____
Client SS#   Client DOB   Representative of JKY Corp.   Date

_____ _____,_____ _____
Client's Address Street and Apt. #   Client's Address City   State   Zip Code

(____)-____-_____.   (____)-____-_____.
Client's Home Phone #   Client's Cell Phone #

Acceptable forms of payment: Check, Money Order, Cash, Visa/MC

Make Payments to: JKY Corporation
Mail to: 1234 Borne Street, Any Town, FL 1234

Each Coach's Responsibility State to State

It is your responsibility to stay informed on your State's ethics codes, privacy laws, duty to report, Federal laws such as HIPAA laws. It's a good idea to consult with your attorney before going into business. You will also want to carry liability insurance. Use the professionals for their respective fields, for example an Attorney for law issues, contracts, forming a corporation, an accountant for tax related issues, an insurance agent for liability insurance, etc. We are in a time of change, so most of these laws are changing. I will give you some websites on the resource page to assist you.

HIPAA LAW: The Standards for Privacy of Individually Identifiable Health Information ("Privacy Rule") establishes, for the first time, a set of national standards for the protection of certain health information. The US Department of Health and Human Services ("HHS") issued the Privacy Rule to implement the requirement of the Health Insurance Portability and Accountability Act of 1996 ("HIPAA"). 1 The Privacy Rule standards address the use and disclosure of individuals' health information—called "protected health information" by organizations subject to the Privacy Rule — called "covered entities," as well as standards for individuals' privacy rights to understand and control how their health information is used. Within HHS, the Office for Civil Rights ("OCR") has responsibility for implementing and enforcing the Privacy Rule with respect to voluntary compliance activities and civil money penalties.

The forms and contracts in this manual are to be used as examples only as a teaching tool. It is strongly recommended that you consult with your attorney before designing any forms or contracts for your new company. Remember that the laws are different state to state and you will need to know which law applies to certain situations.

Notes

Rev. Dr. Kevin T. Coughlin Ph.D. Publication Credits

KTC Publishing Phase IIC Coaching, LLC Amazon.com *My Anger Log and Journal Book* 2017

KTC Publishing Phase IIC Coaching, LLC Amazon.com *My Baby's Journal Book* 2017

KTC Publishing Phase IIC Coaching, LLC Amazon.com *The Master Coach's Life Coach Training Guide 2017*

KTC Publishing Phase IIC Coaching, LLC Amazon.com *Call Center Professional and Ethics* 2017

KTC Publishing Phase IIC Coaching, LLC Amazon.com *The Official Gambling Addiction Recovery Coach's Workbook* 2017

KTC Publishing Phase IIC Coaching, LLC Amazon.com *The Official Gambling Addiction Christian Recovery Coach's Workbook* 2017

KTC Publishing Phase IIC Coaching, LLC Amazon.com *My Daily Pet Journal Book; I Love Cats* 2017

KTC Publishing Phase IIC Coaching, LLC Amazon.com *My Daily Pet Journal Book; I Love Dogs* 2017

KTC Publishing Phase IIC Coaching, LLC Amazon.com *Christian Coaching; Life Recovery Coaching Workbook and Journal 2017* Made the Amazon Top 100 Best-Seller List

KTC Publishing Phase IIC Coaching, LLC Amazon.com *Christian Coaching:* **The Master's Guide to Becoming a Professional Christian Life and Recovery Coach** 2017 Made the Amazon.com Top 100 Best- Seller list

KTC Publishing Phase IIC Coaching, LLC Amazon.com *Addiction Professionals AMA/ APA Guide.* 2017 Made the Amazon.com Top 100 Best-Seller list

KTC Publishing Phase IIC Coaching, LLC Amazon.com *My Monthly Journal; A Roadmap for Change.* 2017 Made the Amazon.com Top 100 Best-Seller list

KTC Publishing Phase IIC Coaching, LLC Amazon.com *My Daily Food Journal; A Daily Food Action Plan for Success.* 2017 Made the Amazon.com Top 100 Best-Seller list

KTC Publishing Phase IIC Coaching, LLC Amazon.com *My Daily Prayer Journal; A Walk with Jesus.* 2017 Made the Amazon.com Top 100 Best-Seller list

KTC Publishing Phase IIC Coaching, LLC Amazon.com *My Monthly Journal Book; a Roadmap for Success.* 2017 Made the Amazon.com Top 100 Best-Seller list

KTC Publishing Phase IIC Coaching, LLC Amazon.com *My Daily Meditation Book.* 2017 Made the Amazon.com Top 100 Best-Seller list

KTC Publishing Phase IIC Coaching, LLC Amazon.com *We Can the Anthology; A Collection of Poetry, A Journey Through Addiction and Recovery* 2017

KTC Publishing Phase IIC Coaching, LLC Amazon.com *Prevention; Teaching Teens to Say No to Drugs.* 2017 Made the Amazon.com Top 100 Best-Seller list.

KTC Publishing Phase IIC Coaching, LLC Amazon.com *Relapse Prevention; Long-Term Sobriety.* 2017 Made the Amazon.com Top 100 Best-Seller list.

KTC Publishing Phase IIC Coaching, LLC Amazon.com *Thirty Days of Thoughts About the Holidays.* 2016 Made the Amazon.com Top 100 Best-Seller list.

KTC Publishing Phase IIC Coaching, LLC Amazon.com *Thirty Days of Thoughts About Christian Recovery and the Holidays.* 2016 Made the Amazon.com Top 100 Best-Seller list.

KTC Publishing Phase IIC Coaching, LLC Amazon.com *Recovery & Life Coaching; The Official Workbook for Coaches and Their Clients*. Co-Author Dr. Cali Estes 2016 #1 Best-Seller Amazon.com Top 100 Best-Seller List

KTC Publishing Phase IIC Coaching, LLC Amazon.com *Addictions: What All Parents Need to Know to Survive the Drug Epidemic.* 2016 Made the Amazon.com Top 100 Best-Seller list.

KTC Publishing Phase IIC Coaching, LLC Amazon.com If You Want What We Have; A Journey Through the Twelve Steps of Recovery Workbook and Manual 2015 Made the Amazon.com Top 100 Best-Seller list.

KTC Publishing Phase IIC Coaching, LLC Amazon.com *In the Sunlight of the Spirit* Workbook and Manual 2015

KTC Publishing Phase IIC Coaching, LLC Amazon.com *We Can; A Collection of Poetry, A Journey Through Addiction and Recovery 2016*

KTC Publishing Phase IIC Coaching, LLC Amazon.com *We Can 2; A Collection of Poetry, A Journey Through Addiction and Recovery 2016*

KTC Publishing Phase IIC Coaching, LLC Amazon.com *We Can 3; A Collection of Poetry, A Journey Through Addiction and Recovery 2016*

Tumbleweeds; Feather Books Poetry Series a Book of Poetry Written by Rev. Kevin T. Coughlin Feather Books England May 2002 (In Memory of DeWitt)

Wayne Independent Newspaper Honesdale, PA

News Eagle, Hawley, PA

Reading Eagle, Reading, PA Berks & Beyond

www.addictsrehab.com

My RecoveryRadio.com Host Kent Paul Sept. 11[Th], 2016 Interview

BBS Radio Poetry reading

Blog Talk Radio - Interviews

The Serenity Show - Interview

Passion Diva Radio- Interview

Recovery Starts Here- Interview

www.sacredearthpartners.com - Interview

The Broken Brain (Blog Talk Radio) - Interview

www.eatingdisorderhope.com

Keys to Recovery Newspaper Beth Dewey CEO

www.keystorecovery.com

All 4 Ur Addiction Recovery Referral Resource Guide Jenny Clark Owner

Tripadvisor.com

MindBodyNetwork

Grieving Behind the Badge Peggy Sweeny Founder

www.theaddictsmom.com

In Recovery Magazine

The Sober World Magazine

The Soberworld.com

Shout My Book

Bookgoodies.com

Goodreads

Book Reader Magazine

Awesomegang.com

www.christiancoaches.com
Hardrock Favorites
Metalwork
SixxAMFansPhotos
NEWS CHANNEL 10 EYEWITNESS NEWS CHANNEL.COM
KHQQ6 ABC NEWS
ABC EYEWITNESS NEWS 8 KLKN-TV
FOX14 NEWS AT 9
Erie News Now
NTV Nebraska.TV ABC
Western Mass News Channels 3 ABC 40 Fox 6
ABC9 KTRE
7 KLTV ABC
Fox 19 Now
KXNEWC Eyewitness News
12 WSFA ABC
ABC 6 News WLNE TV
100.7 KFM BFM
Fox 5 KVVU-TV Local Los Vegas
13 WTHR COM Indians News ABC
Eyewitness News 3 WFSB.COM
Fox 12 Oregon
WDRB.COM
Fox29 WFFX.COM
WETV San Diego
HAWAII News Now
Marketers Media
WALB News 10 ABC
Tristate Update.com 13 News WOWK
AM760
WMBF ABC News
KCEN HD ABC KCENTV.COM
WECT6 ABC News
Eyewitness News3 WFSB.COM
WLOX ABC BOUNCE Eyewitness News
Eyewitness News 8
CBS8.COM
News channel 6 KAUZ
SPROUT News
12 Eyewitness News KFVS
KEYC MANKATO News 12 CBS & FOX LOCAL NEWS
3 WRCB TV ABC COM
KNDO 23 NBC
KNDU 25 NBC
RecoveryView.com
The Aurorean, Encircle Publications 1998 Poetry and Essays

Joel's House Publications 1998-2005 Poetry and Essays
Our Journey 1998-2005 Poetry
The Poetry Explosion, The Pen 1999-2003 Poetry
Apostrophe 1998 Poetry
Nuthouse Twin Rivers Press 1998 Poetry
The National Library of Poetry 1998
Lines N' Rhymes 1998 Poetry
The Poetry Church Feather Books
England. Anthology John Hunt Publications 1999 Poetry
A Tapestry in Time. 1999 Poetry Book 18 Poems
Connecticut Department of Mental Health and Addiction Services
The Webster Times 1999 Poetry
The Angel News 1999 Poetry
The Skater won The Editor's Choice Award September 1999 (Our Journey)
The Blind Man's Rainbow 1999 Poetry
Anabella 2001 Poetry
Feather Books, The Poetry Church 1998-2002
The American Dissident 2002 Poetry
The Good Shepherd Poetry 2002
Yak ' Suo Magazine Essays and Poetry
Colt. Winner Editor's Choice Award Contest Literally Horses 2002
Goodbye My Friend Read on the Radio Rhyme and Reason UBC Europe & the UK September 2001 Read on the Radio in Europe and the UK as a Tribute to those lost on September 11th bombings. My poem was read over the radio for many days.
Tumbleweed Read on BBC Radio in England 2001
Published by Feather Books
Notified by John Waddington Feather that Tumbleweed had been read on BBC Radio in England on Several Occasions.
Stanwick Congregational Flyer Poetry
University of Scranton Aniska College of Professionals Essay 2002
Scranton University 2002 Poetry
The River Reporter Newspaper 2002 Poetry
Unity Community News 2002 Poetry
The Poetry Corner Angelfire.com Poetry
The Poet's Market 2002 Poetry
The Poetry Church England 2003 Poetry
Cover of Wayne Independent News 2003 Poetry
Nomad's Choir 2003 Poetry
Written a series of 9 course manuals for a coaching recovery curriculum. 2014-2015
www.addictedminds.org 2015-2016 Articles
www.soberservices.com 2015 Articles
http://fromaddict2advocate.blogspot 2016 Articles Marilyn Davis
LinkedIn 2014-2016 Articles
Two Drops of Ink S.W. Bedful 2015- 2016 Poetry/ Articles
The Addict's Mom 2016 Articles Blog
Ghostwriter Articles/ Content 2014-2016

KEITV12: The Kingdom Hour- Interview
BlogTalkRadio The Kingdom Hour- Interview
RecoveryView.com Online Journal Interview
Brooklyn TV Interview
CCE Christian Children's Empowerment

About the Author

Rev. Dr., Kevin T. Coughlin Ph.D., DCC, DDV, DD, IMAC, NCIP is an International Master Coach, trainer, best-selling author, writer, poet, speaker, a Diplomate Christian counselor, and therapist, he is Board Certified in Family, Developmental, Alcoholism, Substance Abuse, and Grief Counseling, the Reverend is a NCIP interventionist, a Domestic Violence Advocate, Associate Professor for DCU, a Provincial Superintendent (to be consecrated a Bishop in 2018) and so much more; he is an expert in the field of Addiction and Recovery. He is also a Founder of a Residential Recovery Facility New Beginning Ministry, Inc. and President and CEO of Phase IIC Coaching, LLC., and was The Program Director for a Professional Coaching Training School, and was the Editor in Chief for an online website and journal.

The Reverend has over forty-seven years of experience with the AA program. He has been working in the addiction recovery field for over two decades, has helped thousands of individuals and their families overcome all types of addictions, substance abuse, alcoholism, process addiction, shame and guilt, relationship and communication problems, anger management, inner healing, self-image, interventions and much more. He is a Best-Selling author and Award-Winning Poet and has published articles throughout the United States and other Nations, he has been interviewed on numerous radio talk shows, television, published in magazines, newspapers, books, and online publications; he has been featured on ABC, CBS, FOX, NBC, and the BBC in the UK. Rev. Kev is a former State, National & World-Champion Powerlifter, and still, holds several records. He loves to write, read, teach, listen to music, and spend time with people and dogs. His parents are his heroes.

Follow Rev. Kev. on Social Media

https://www.goodreads.com/author/show/14874631.Kevin_Coughlin

About Me Link: https://about.me/ktc1961/

http://ilikeebooks.com/if-you-want-what-we-have/

http://awesomegang.com

www.amazon.com/Rev.-Kevin-TCoughlin/e/B01AF6AAAI/ref=ntt_dp_epwbk_0

http://www.barnesandnoble.com/w/addictions-what-all-parents-need-to-know-to-survive-the-drug-epidemic-rev-dr-kevin-t-coughlin-phd/1124049106?ean=9780997700695

http://www.barnesandnoble.com/w/in-the-sunlight-of-the-spirit-rev-dr-kevin-t-coughlin/1124049139?ean=9780997700671

http://www.barnesandnoble.com/w/if-you-want-what-we-have-rev-dr-kevin-t-coughlin/1124049130?ean=9780997700688

http://mybookplace.net/in-the-sunlight-of-the-spirit-a Rev. Dr. Kev's Social Media Accounts

Facebook
1. Kevin Coughlin: https://www.facebook.com/profile.php?id=100008449955607
2. My Group, Resources for those suffering from addiction and their families: https://www.facebook.com/groups/resourcesforthosesufferingfromaddiction/
3. RevKev The Addiction Expert: https://www.facebook.com/RevKev/?fref=ts

LinkedIn
1. Rev. Dr. Kevin T. Coughlin PhD
   https://www.linkedin.com/in/revkevnetwork

Google+
1. Kevin Coughlin
   https://plus.google.com/112400908736308001821/posts
   My Group: The Recovery Community Family and Friends:
   https://plus.google.com/communities/113521225141112811207

Pinterest
1. Kevin Coughlin: https://www.pinterest.com/ktc1961/
2. My Group Board: Recovery We Can
   https://www.pinterest.com/ktc1961/recovery-we-can/

Tumblr
1. https://www.tumblr.com/blog/revkevsrecoveryworld

Instagram
theaddiction.expert

My Websites:
1. www.revkevsrecoveryworld.com
2. theaddiction.expert
3. theaddiction.guru

Twitter:
1. https://twitter.com/AuthorRevKev
Rev. Kev's Goodreads Link:
-spirituality-training-manual-and-workbook-by-kevin-coughlin

Friends of Recovery Readers and Reviewers Book Group on Facebook:
https://www.facebook.com/groups/1667708970205824/

Thank you for reading my work! If you enjoyed my book, would you consider reviewing it on Amazon.com? We would appreciate your help in getting the word out on how helpful this book can be in someone's life. Thank you so much and God bless you! Phil 4:13